Asphyxia in Neo

In severe asphyxia, lipid peroxidation occurs
concentration may indicate the severity of asp
reason for needing cooling therapy. This bool
results of research into the impact of oxygen c
neonates. The physiological consequences of a
newborn infants include changes in the pH of
lipid concentrations. These changes can result i
loss of cell membrane integrity, and damage to

Key Features

- Reviews pathophysiology of asphyxia
- Describes the influence of asphyxia on lipid concentrations
- Summarizes the weight distribution of neonat asphyxia based on lipid concentration
- Illustrates the use of lipids as an indicator of p following asphyxia
- Provides guidance for lipid parameters in cooli

Asphyxia in Neonates

Miljana Z. Jovandaric

with contributions by
Stefan Dugalic and Sandra Babic

CRC Press
Taylor & Francis Group
Boca Raton London New York

CRC Press is an imprint of the
Taylor & Francis Group, an **informa** business

Designed cover image: MD. MUNNA ISLAM

First edition published 2024
by CRC Press
2385 NW Executive Center Drive, Suite 320, Boca Raton FL 33431

and by CRC Press
4 Park Square, Milton Park, Abingdon, Oxon, OX14 4RN

CRC Press is an imprint of Taylor & Francis Group, LLC

ISBN: 978-1-032-38771-0 (hbk)
ISBN: 978-1-032-38772-7 (pbk)
ISBN: 978-1-003-34669-2 (ebk)

DOI: 10.1201/9781003346692

Typeset in Garamond
by SPi Technologies India Pvt Ltd (Straive)

Contents

About the Author

Miljana Z. Jovandaric, MD, PhD, was born in Serbia. She graduated from the Faculty of Medicine, University of Belgrade, and completed her pediatric specialization in 1999 at the University Children's Hospital, Belgrade. Dr. Jovandaric completed her specialization in neonatology in 2003. She defended her master's thesis, "Analysis of lipid infants in women suffering from gestational diabetes mellitus (GDM)," in 2006 and doctoral dissertation, "Effect of hypoxia on electrolyte and lipid levels in term newborns," in 2018 as part of the medical faculty at University of Belgrade, Serbia. Dr. Jovandaric has been a research associate since 2020 as part of the medical faculty of the University of Belgrade, Serbia. Author and coauthor of 106 scientific papers presented at national and international conferences and journals, she's head of the Department of Sick Newborns at the Clinic for Gynecology and Obstetrics, Department of Neonatology, University Clinical Center of Serbia, Belgrade. Dr. Jovandaric has edited and/or contributed to the following books: *Childbirth: Placenta Abruption and Delivery Method*; *Selected Topics in Myasthenia Gravis: Maternal and Neonatal Outcome of Pregnancies with Autoimmune Myasthenia Gravis*; *Apoliproteins, Triglicerides and Cholesterol: Significance of Lipid and Lipoprotein in Organism*; and *Veganism: A Fashion Trend or Food as a Medicine – Veganism: A New Approach to Health.*

About the Contributors

Stefan Dugalic, Ass., MD, PhD, is a gynecology and obstetrics specialist, and master manager in the medical/health care system. In 2017, he joined the Clinic for Gynecology and Obstetrics, University Clinical Centre of Serbia, as doctor of medicine, and later as a gynecology and obstetrics specialist in 2021. During 2019, he participated in the project of the Ministry of Education, Science and Technology developing clinical epidemiological research of health disorders of public health significance for the population of Serbia (project No. 175025). Dr. Dugalic received his MD from the School of Medicine at the University of Belgrade in 2016, where he also completed his education as a specialist in the field of human reproduction in 2017 and in the field of gynecology and obstetrics in 2021.

Sandra Babic, MD, was born in Belgrade, Serbia, and graduated from the Medical Faculty of the University of Belgrade. At the same faculty in 2015/2016, she enrolled and completed the specialist academic studies program in the field of human reproduction. She completed her study as a specialist in gynecology and obstetrics in 2020. She has been with the Clinic for Gynecology and Obstetrics UKCS since 2013, first as a volunteer and now as a clinical doctor. Dr. Babic has published several works in foreign and domestic professional journals.

Preface

Asphyxia in Neonates is a product of my long-term work with asphyctic newborns. This topic was the inspiration for my master's thesis, as well as doctoral studies and papers published in numerous journals. During the perinatal period, asphyxia is defined as a state of disturbed gas exchange, which is accompanied by an increase in lactate, carbon dioxide, and other waste products, as well as a significant decrease in the concentration of oxygen in the blood that leads to lipid peroxidation, damage and permeability of capillaries, and complex metabolic processes, and causing subsequent damage to the heart, brain, kidneys, and other organs. For years, different treatment modalities have been applied, as well as parameters for sending patients to therapeutic hypothermia. Lipid peroxidation and determining the concentration of lipids in the serum of asphyctic children proved to be a good predictive indicator of the severity of asphyxia.

Miljana Z. Jovandaric, MD, PhD

Chapter 1

Perinatal Asphyxia, Hypoxia, Ischemia, and the Newborn

Miljana Z. Jovandaric

1.1 Introduction

Perinatal asphyxia is a complex clinical problem that is most often defined as a disorder of respiratory gas exchange in the placenta of the fetus during delivery or in the lungs of the newborn after birth (1). Although the word "asphyxia" itself means "without a pulse" (Greek: *a* – no; *sphygmos* – beating pulse, suffocation, apparent death), asphyxia in the clinical sense means the condition of a newborn after birth, accompanied by cessation of breathing (apnea) with preserved or weakened pulse. The term *asphyxia* in medical terminology represents a state of neonatal respiratory distress with short-term and long-term effects on the body. Asphyxia can also be defined as a state of impaired gas exchange, accompanied by an increase in lactate, carbon dioxide, and other waste

DOI: 10.1201/9781003346692-1

products, as well as a significant decrease in the concentration of oxygen in the blood (2).

While the term *asphyxia* is reserved for the combined state of hypoxia, hypercapnia, and acidosis, hypoxia (Greek *hypo* + Latin *oxygen* + Greek *haima* – blood) refers to states of oxygen deficiency without accompanying acidosis. Hypoxia means reduced tissue oxygen supply and is a consequence of reduced oxygen pressure in the blood (hypoxemic hypoxia), reduced blood flow (ischemic hypoxia), or reduced capacity of the blood to supply oxygen (anemic hypoxia). Ischemia is a decrease in blood flow, with a consequent decrease in oxygen and glucose in the required amounts, which leads to organ damage (3).

Hypoxia as a physiological-pathophysiological event has an important meaning during normal and pathological childbirth. During the initial dilation of the cervix, the pH of the fetal blood drops from 7.37 to 7.27 at full dilation. A certain degree of hypoxia and acidosis is present during normal labor. The variability of gases in the arterial blood is mild and reversible, so it is believed that this "transient asphyxia" has a physiological role in preparing the fetus for the beginning of normal breathing (4).

Most children (more than 95%) overcome this condition without major problems. Only a small number require active assistance during the first breath and the beginning of alveolar ventilation (which is actually a condition for survival). If the initiation of breathing is unsuccessful, respiratory stimulation will quickly prevent pathological inversion of blood gases in the newborn and establish extracellular and intracellular homeostasis. Thus, primary resuscitation by shortening the period of hypoxia, hypercapnia, and acidosis will prevent perinatal hypoxic tissue damage, the most severe of which is brain damage. Because of all this, asphyxia should be understood as a medical term for a clinical manifestation (the child is not breathing!), which is a consequence of varying degrees of hypoxia (5).

Although the pathophysiological mechanisms of severe asphyxia, hypoxia, or ischemia (focal or global) can be specifically simulated and investigated in research on animal models, in clinical reality it is almost impossible to separate these conditions, so they are labeled as hypoxic-ischemic damage. In human newborns, it is often very difficult to determine the duration of hypoxia and its intensity during the birth itself. There is also an attempt to measure the effect of anoxia (hypoxia) by the degree of asphyxia in a newborn whose potential is only partially known (e.g., gestational age) and whose resistance or sensitivity to different degrees of hypoxia are different from other neonates (6). The complexity of the definition of asphyxia/hypoxia/ischemia is best reflected in the recommendations of numerous authors for the more precise diagnosis of asphyxia/hypoxia by connecting clinical signs to biochemical parameters. The previous term *hypoxic-ischemic encephalopathy* has become insufficiently precise. Since it is not only about hypoxia and ischemia, a more adequate term based on current knowledge is *hypoxic-ischemic-reperfusion-reoxygenation brain lesion* (7).

1.2 Perinatal Asphyxia and the Newborn

There is a lack of data on the incidence of perinatal asphyxia/hypoxia. In developed countries with well-organized perinatal services, the incidence is 1–6/1,000 term newborns (8). The assessment of the incidence of asphyxia varies due to uneven attitudes, depending on whether only full-term, premature, or all children with signs of asphyxia are taken into account. In countries with a low level of development, the incidence ranges up to 26/1,000 live births. It is estimated that 30% of children have hypoxia/ischemia in developed countries, and 60% in developing countries have some evidence of intrapartum hypoxia/ischemia (9).

One of the most common causes of asphyxia is premature birth, with all the consequences that accompany this group of newborns. Premature birth is defined as any birth at a gestation period of less than 37 weeks of gestation. Extremely preterm is less than 28 weeks (10).

Premature birth is the most common cause of infant death worldwide. About 15 million babies are born prematurely each year (5% to 18% of all births). Late preterm birth accounts for 75% of all preterm births. This rate varies between countries. In the United Kingdom, 7.9% of babies are born preterm, and in the United States, 12.3% of all births are before 37 weeks of gestation (11). About 0.5% of births are extremely early births (20–25 weeks of gestation), and they account for the majority of deaths. In many countries, the rate of preterm birth increased between the 1990s and 2010s. Complications from premature birth led to 810,000 deaths in 2015 worldwide, compared to 1.57 million in 1990 (12).

The chance of survival at 22 weeks is about 6%, while at 23 weeks it is 26%, at 24 weeks 55%, and at 25 weeks about 72%. Chances of survival without long-term difficulties are less. Prematurity has increased in prevalence over the past few decades in almost all countries with reliable data on the trend. In 2014, preterm birth had a worldwide prevalence of 10.6%, affecting nearly 15 million births annually. Due to advances in treatment introduced during the 1970s and 1980s, the survival of preterm infants has also improved dramatically (13). Over 95% of preterm infants and most of those born extremely preterm (<28 weeks) who receive modern neonatal and pediatric care now survive (14).

Premature birth is one of the biggest public health problems of recent decades. The increased percentage of survival represents a great challenge for perinatal medicine to understand and prevent the neurodevelopmental consequences in a premature child. The first neonatal intensive care unit was established in the United States in 1965 at the Yale University

School of Medicine. Exogenous surfactant was used for the first time in the treatment of respiratory distress syndrome in 1980 (15). The next advance in perinatal medicine is the antenatal use of steroids, after the New York Declaration of 1994, formalized in the National Institute of Health consensus statement in 1994 and recognized by the American College of Obstetricians and Gynecologists soon after. The use of surfactants and antenatal corticosteroids are revolutionary therapies that have increased the survival of premature birth in children, but the long-term neurodevelopmental consequences are still being explored (16).

The application of new knowledge in clinical practice as well as the development of technology in the form of the use of ventilators with different modes of ventilatory support, adequate resuscitation in the first minutes of life, combined with the knowledge of the nutritional needs of newborns in the last 50 years, have significantly improved the outcomes and consequences for children who were born prematurely. Complications caused by incomplete growth and development of organs can be mitigated by adequate therapy in intensive care units. Adequate therapy and care in the first minutes, hours, days, and weeks as well as a multidisciplinary approach by a team of experts from various specialties are key factors for a favorable outcome and further progress of premature children. During the long-term follow-up of prematurely born children, various complications can be observed that can affect the quality of life. Prematurely born children often faces increased risks of psychological, neurological, and physical diseases. Adequate recognition of problems in primary and secondary health care institutions can prevent or stop the progression of the disease (17).

Extensive clinical and research data support the fact that being born prematurely is associated with an increased risk of long-term health and neurodevelopmental problems. Many factors have been studied that could have an impact, primarily neurodevelopment. Environmental factors have an impact on brain maturation in the first days of life. The benefit of mother's milk and early breastfeeding is important in order to prevent necrotic enterocolitis, but the benefit for neurodevelopment has not been clearly proven. It has been established that noise, pain, and the environment of neonatal intensive care can negatively affect the neurodevelopment of newborns. Isolated intensive care rooms have no advantage over open rooms. The positive effects of the presence of parents with newborns (rooming-in) on shaping the long-term neurodevelopment of high-risk premature babies have been proven (18).

1.3 Division of Asphyxia (Hypoxia) According to Duration

1.3.1 Acute Asphyxia

The condition of acute asphyxia has been well studied in animal models. After a few superficial breathing movements in which there is no gas exchange, newborn animals stop breathing, which is called primary apnea and can last up to 10 minutes. However, most animals start gasping, increasing the frequency and power of breathing, which then gradually decrease until the last breath. This period of gasping is followed again by apnea, but this time it is terminal apnea (19).

Heart rate drops rapidly after birth, increases in primary apnea and early gasping, and then slows. Heart activity continues for 10 minutes or more after the last breath. The period from the last breath to the cessation of cardiac activity is called secondary or terminal apnea. At the end of terminal apnea, severe metabolic or mixed acidosis develops: pCO_2 exceeds 100 mmHg, pH < 6.5, and pO_2 is unmeasurable. Hyperkalemia develops with $K^+ > 15$ mmol/L. Newborn primates can survive a complete lack of oxygen for 20 minutes. This is made possible by large reserves of glycogen in the brain, liver, and myocardium, which can produce energy anaerobically, as well as the ability of neonatal tissues to metabolize lactates and ketones (20).

The animal's reaction to asphyxia depends on the stage of asphyxia. If the animal receives oxygen or air after terminal apnea, it no longer initiates spontaneous breathing. If only asphyxia is observed in the human newborn, the response depends on the pH of the blood at the time of birth. If its value is higher than 7.25 (primary apnea, mild acidosis), it starts breathing spontaneously. If its value is 7.00–7.10 (moderate acidosis), it can still breathe spontaneously but requires monitoring in the postpartum period. In very severe intrapartum asphyxia (pH < 7.00 [severe acidosis]), the child is usually lethargic, with bradycardia and terminal apnea. Reanimation of the newborn is required. After resuscitation, a certain number of newborns make a quick and complete recovery. In the majority, however, morphological and functional damage to numerous organ systems occurs, known as "post-asphyxia syndrome" (21).

1.3.2 Chronic Asphyxia

In chronic asphyxia, the fetus suffers from occasional repeated episodes of asphyxia (uncoordinated uterine contractions or occasional episodes of pressure on the umbilical cord), with a period of short-term recovery between them. During these episodes, blood pressure increases in the fetus, there is bradycardia, and fetal pO_2 (partial arterial oxygen pressure) is lowered. The process of glycolysis involves the breakdown of glucose via pyruvate to lactate under anaerobic conditions. pCO_2 (partial pressure of carbon dioxide) increases, and usually as a result, mixed respiratory and metabolic acidosis occurs. If such a child is born immediately, their breathing is suppressed, but they recover very quickly after resuscitation with no significant consequences (that is, if there are no neonatal complications). However, if these episodes last longer, the fetus fails to recover between individual interruptions in oxygen supply. Hemodynamic compensatory possibilities are gradually exhausted, and hypotension and deeper acidosis occur. After delivery and despite rapid resuscitation, such a child will have early and late signs of post-asphyxia (22).

1.3.3 Division of Asphyxia (Hypoxia) Based on Apgar Score

Apgar represents a summary assessment of the vital parameters of a newborn. It is named after Virginia Apgar, an American anesthesiologist, who in 1953 introduced this simple, systematic examination of newborns, which includes five parameters: skin color, respiration, heart rate, muscle tone, and response to stimulation. Each of the parameters is evaluated with a score of 0, 1, or 2. The total score is the Apgar score (AS). A favorable Apgar is 8–10, moderate depression is indicated by AS 4–7, and severe depression by AS 0–3. Apgar is evaluated in the first (vitality) and fifth (adaptation) minutes after birth (23).

A low Apgar (<7) indicates the need for resuscitation, and the assessment can be performed further, thus monitoring the success of resuscitation. Apgar cannot be used to assess long-term prognosis. For some time, neonatology associations have been talking about the possibility of introducing a new, more precise AS. Specifically, skin color is an unreliable clinical parameter of vitality, especially in the first minute when the expected oxygen saturation is only 60–70%. At this saturation, the skin cannot be clearly pink, for which a score of 2 is given, but is slightly cyanotic (with a bluish tinge), resulting in a lower score. Oxygen saturation reaches a value of around 90% only around the fifth minute, which is when the skin becomes really pink (24–26).

1.3.4 Room Air in Newborn Resuscitation

Hyperoxia has an important effect on circulation, especially in the brain and lungs. Although research results are still not clear enough, there is evidence that hyperoxia can cause vasoconstriction in the brain. The same effect can occur in the pulmonary circulation, exacerbating pulmonary hypertension. A study of a group of newborns with circulatory collapse showed that circulatory recovery is better if 21% oxygen is used during resuscitation instead of 100% (27–29). Since 2001, room air began to be used in the resuscitation of newborns, but still without official guidelines from the competent medical associations (International Liaison Committee on Resuscitation [ILCOR], American Association of Pediatrics [AAP], among others). The reason for delaying the official introduction of room air as the first choice in newborn resuscitation was an unresolved question: How to act if the saturation during resuscitation with room air remains significantly lower than normal values? Consequently, the ILCOR/AAP guideline from 2005 reads, "The optimal concentration of oxygen for resuscitation of newborns is not known." Based on this, neonatologists around the world are free to exercise their own discretion whether to use oxygen concentrations from 21% to 100%. An increased concentration of oxygen was applied if the saturation was lowered, and soon the same was done if the heart rate dropped in the first 60–90 seconds of life or did not increase to the expected 110 beats/minute. The next step was to determine the saturation reference values in the first minutes of life. According to the European Resuscitation Council's life support algorithm, the application of a concentration of oxygen higher than 21% is warranted if the saturation is lower than the following values: 60%, 70%, 80%, 85%, and 90% in the 2nd, 3rd, 4th, 5th, and 10th minute of life, respectively (30, 31). In infants aged >32 weeks of gestation (GW), it is best to start resuscitation with room air.

If the heart rate drops in the first 90 seconds of life, apply oxygen, and in infants aged <32 GW, start resuscitation with 21–30% oxygen. If the saturation is <70% in the fifth minute and/or the heart rate is decreased, apply 100% oxygen. For newborns with extremely low birth weight (<1,000 g), the target saturation in the postnatal period should be 89–94%. Research into the optimal oxygen concentration for newborns continues. The tendency of the current guidelines is to limit the use of 100% oxygen. In addition to guidelines for resuscitation, an answer should be given to the question of oxygen administration during all 30 days of the neonatal period. There are authors who rightly ask the question of adequate oxygen concentration for older children. Guidelines for resuscitation and oxygen concentration of extremely premature infants (<27 GW) have not yet been specified (32).

1.4 Neonatal Resuscitation in Relation to Apgar Score

An Apgar score of 4–7 is a sign of moderate, and 3–4 of severe, asphyxia, which usually requires resuscitation with positive air pressure and oxygen as needed. An Apgar score of 0, 1, or 2 is a sign of the most severe form of asphyxia that requires complete resuscitation (usually intubation, application of positive air pressure and oxygen, and medications such as adrenaline, bolus 0.9% NaCl, and inotropes) (33).

1.5 Clinical Presentation of Asphyxia/Hypoxia in Newborns

Despite the monitoring of vital parameters before delivery, it is often difficult to determine the beginning (during delivery or before delivery) and the duration of the asphyxia episode. If the interruption of oxygen supply is acute, the clinical picture is expressed by a very low Apgar score. If the lack of oxygen is chronic, there is an adaptation of the fetal circulation to increasingly strong hypoxemia by centralizing the blood flow, so that the symptoms of asphyxia (hypoxia) at birth can be latent, and the real clinical picture of hypoxia occurs later, i.e., usually a few hours after birth due to dysfunction of certain organs or, more often, organ systems (lungs, heart, brain, intestines, kidneys) (34).

Perinatal asphyxia is an emergency, which is why an adequate and timely therapeutic procedure is of crucial importance. Examination of biochemical processes during asphyxia/hypoxia in animal model experiments contributed to the introduction of protocols/algorithms for resuscitation. The basis of the therapeutic procedure is aspiration of the airways and resuscitation with room air, i.e., controlled application of oxygen (monitored by a pulse oximeter). Although cardiorespiratory symptoms are the most noticeable, there is the involvement of a number of organs, and the irreversible consequences most often affect the central nervous system (35).

1.6 Pathophysiology of Newborn Asphyxia

In conditions of hypoxia and ischemia due to reduced production of adenosine triphosphate (ATP), many structural and functional changes occur, especially in highly differentiated cells of the CNS, heart, and kidneys. It is certain that loss of cell membrane function is the primary event in the occurrence of cell damage under hypoxic conditions (1). Numerous experimental studies support the importance of free radical production after acute hypoxia. All metabolic changes associated with reversible changes in cell hypoxia, if qualitatively or quantitatively significant, can cause irreversible damage (36).

In hypoxia, the tissue that was in normal conditions in a dynamic equilibrium on the one hand with contact with substances, and on the other with removal of waste products, now turns into a closed system. Oxidative phosphorylation and metabolism of oxygen-dependent substances stop, metabolic changes occur depending on internal reserves that are quickly consumed, and metabolic end products harmful to the tissue accumulate, which leads to the extinction and destruction of cells and tissues (37).

In hypoxia, compared to the normal state of oxygen exposure, the life conditions of the fetus change. Acute or chronic suffering of the fetus occurs. Fetal hypoxia can occur as a result of disturbed oxygen transport at various stages: maternal hypoxia, reduction of uterine blood flow, disruption of placental function, and changes in the umbilical cord. In the placenta, the microcirculation is disturbed, which causes a metabolic disorder in the mother–fetus system, because there is a disturbance in the exchange of nutrients and, later, in the transport of oxygen. Fetal hypoxia occupies an important place in the causes of postnatal mortality and somatic and neurological disorders of children. Hypoxia is the primary cause of various consequent pathological conditions. The essence of the hypoxic process is an energy deficit, which leads to functional and morphological changes and ultimately to cell death (38).

Hypoxia disrupts cellular metabolism in many different ways. The limited reserve of glycogen plays a central role. Under aerobic conditions, glycolysis (via the Krebs cycle and oxidative phosphorylation) produces 38 molecules of ATP from 1 molecule of glucose. ATP powers the neuronal membrane pump that keeps the internal concentration of potassium high and sodium, calcium, and chloride low. This ionic balance is the physiological basis for intraneuronal communication. With the lack of glucose and oxygen, all ATP-dependent processes are interrupted. In complete hypoxia/ischemia, the cell consumes all glycogen and all ATP stores within 10 minutes. Anaerobic metabolism begins in the cell (it produces only two molecules of ATP from each molecule of glucose), which leads to a decrease in pH. Without ATP, the ion pump loses function, sodium and chlorine enter the cell, and then water follows by passive diffusion with the ultimate effect of cytotoxic cellular edema. The cell swells, ruptures, and releases its cytoplasmic contents into the interstitial space (39).

1.7 Conclusion

Regardless of which term is used, asphyxia, hypoxia, or ischemia in the newborn age implies insufficient oxygenation of tissues and organs with consequential damages that can be manifested in different periods of time and depend on the duration of insufficient oxygenation.

Abbreviations

AS Apgar score
ATP adenosine triphosphate
CNS central nervous system
GW gestational week

References

1. Li C, Miao JK, Xu Y, Hua YY, Ma Q, Zhou LL, Liu HJ, Chen QX. Prenatal, perinatal and neonatal risk factors for perinatal arterial ischaemic stroke: a systematic review and meta-analysis. *Eur J Neurol.* 2017;24:1006–1015.

2. Carter BS, Havercamp AD, Merenstein GB. The definition of acute perinatal asphyxia. *Clin Perinatol.* 1993;20:287–303.
3. González de Dios J. Definition of perinatalasphyxia in medical literature: the need of a consensus. *Rev Neurol.* 2002;35:628–634.
4. Hill A, Volpe JJ. Perinatal asphyxia: clinical Aspects. *Clin Perinatol.* 1989;16:435–454.
5. Simon NP. Long-term neurodevelopmental outcome of asphyxiated newborns. *Clin Perinatol.* 1999;26:767–778.
6. Golubnitschaja O, Yeghiazaryan K, Cebioglu M, Morelli M, Herrera-Marschitz M. Birth asphyxia as the major complication in newborns: moving towards improved individual outcomes by prediction, targeted prevention and tailored medical care. *EPMA J.* 2011;2:197–210.
7. González de Dios J, Moya M, Vioque J. Risk factors predictive of neurological sequelae in term newborn infants with perinatal asphyxia. *Rev Neurol.* 2001;32:210–216.
8. Ferriero DM Neonatal brain injury. *N Engl J Med.* 2004;351:1985–1995.
9. Kurinczuk JJ, White-Koning M, Badawi N. Epidemiology of neonatal encephalopathy and hypoxic-ischaemic encephalopathy. *Early Hum Dev.* 2010;86:329–338.
10. March of Dimes, PMNCH, Save the Children, WHO. *Born too soon: the global action report on preterm birth.* Howson CP, Kinney MV, Lawn JE, editors. Geneva: World Health Organization; 2012.
11. Chawanpaiboon S, Vogel JP, Moller AB, Lumbiganon P, Petzold M, Hogan D. Global, regional, and national estimates of levels of preterm birth in 2014: a systematic review and modelling analysis. *Lancet Glob Health.* 2019;7:e37–e46.
12. Manley BJ, Doyle LW, Davies MW, Davis PG. Fifty years in neonatology. *J Paediatr Child Health.* 2015;51:118–121.
13. Raju TNK, Buist AS, Blaisdell CJ, Moxey-Mims M, Saigal S. Adults born preterm: a review of general health and system-specific outcomes. *Acta Paediatr.* 2017;106:1409–1437.
14. Raju TNK, Pemberton VL, Saigal S, Blaisdell CJ, Moxey-Mims M, Buist S. Long-term healthcare outcomes of preterm birth: an executive summary of a conference sponsored by the National Institutes of Health. *J Pediatr.* 2017;181:309–318. (e1).

15. Fujiwara T, Adams FH. Surfactant for hyaline membrane disease. *Pediatrics.* 1980;66 (5):795–798.
16. Myers E, Ment LR. Long-term outcome of preterm infants and the role of neuroimaging. *Clin Perinatol.* 2009;36 (4):773–789.
17. Crump C, Sundquist J, Winkleby MA, Sundquist K. Preterm birth and risk of chronic kidney disease from childhood into mid-adulthood: national cohort study. *BMJ.* 2019, 365:l1346.
18. Cheong JLY, Burnett AC, Treyvaud K, Spittle AJ. Early environment and long-term outcomes of preterm infants. *J Neural Transm (Vienna).* 2020;127:1–8.
19. Gill MB, Perez-Polo JR. Hypoxia ischemia-mediated cell death in neonatal rat brain. *Neurochem Res.* 2008;33:2379–2389.
20. Vannucci RC, Vannucci SJ. Perinatalhypoxic-ischemicbrain damage: evolution of an animal model. *Dev Neurosci.* 2005;27:81–86.
21. Ahearne CE, Boylan GB, Murray DM. Short and long term prognosis in perinatal asphyxia: an update. *World J Clin Pediatr.* 2016;5:67.
22. Jensen A, Garnier Y, Berger R. Dynamics of fetal circulatory responses to hypoxia and asphyxia. *Eur J Obstet Gynecol Reprod Biol.* 1999;84:155–172.
23. Apgar V. A proposal for a newmethod of evaluation of the newborn infant. *Curr Res Anesth Analg.* 1953;32:260–267.
24. Finster M, Wood M. The Apgar score has survived the test of time. *Anesthesiology.* 2005;102:855–857.
25. Vento M, Asensi M,Sastre J,Garcia-Sala F, Pallardo FV,Vina J. Resuscitation with room air instead of oxygen prevent so xidative stress in moderately asphyxiated term newborns. *Pediatrics.* 2001;107:642–647.
26. Vento M, Saugstad OD. Resuscitation of the term and preterm infant. *Semin Fetal Neonatal Med.* 2010;15:216–222.
27. Liu JQ, Saugstad OD, Cheung PY. Using 100% oxygen for the resuscitation of term neonates until evidence of spontaneous circulation: more investigations needed. *Resuscitation.* 2010; 81:145–147.
28. Solas AB, Kalous P, Saugstad OD. Reoxygenation with 100% or 21% oxygen after cerebral hypoxemia-ischemia hypercapnia in newborn piglets. *Biol Neonate.* 2004;5:105–111
29. Lapointe A, Barrington KJ. Pulmonary hypertension and the asphyxiated newborn. *J Pediatr.* 2011;158 (2 Suppl):e19–e24.

30. Dawson JA, Vento M, Finer NN, Rich W, Saugstad OD, Morley CJ, Davis PG. Managing oxygen therapy during delivery room stabilization of preterm infants. *J Pediatr.* 2012;160:158–161.
31. Villani P, Di Fabio S, Torielli F, Gizzi C, Martano C. Introduction to the principles of neonatal resuscitation. *Minerva Pediatr.* 2010;62 (3 Suppl 1):225–226.
32. Biban P, Filipovic-Grcic B, Biarent D, Manzoni P, International Liaison Committee on Resuscitation (ILCOR); European Resuscitation Council (ERC); American Heart Association (AHA); American Academy of Pediatrics (AAP). New cardiopulmonary resuscitation guidelines 2010: managing the newly born in delivery room. *Early Hum Dev.* 2011;87 (Suppl 1):S9–11.
33. Berger TM, Pilgrim S. Resuscitation of newborn infants. *Anaesthesist.* 2009;58:39–50.
34. Fattuoni C, Palmas F, Noto A, Fanos V, Barberini L. Perinatal asphyxia: a review from a metabolomics perspective. *Molecules.* 2015;20:7000–7016.
35. Shah P, Riphagen S, Beyene J, Perlman M. Multiorgan dysfunction in infants with post-asphyxial hypoxic-ischemic encephalopathy. *Arch Dis Child Fetal and Neonatal.* 2004;89:152.
36. Vannucci RC, Towfighi J, Vannucci SJ. Secondary energy failure after cerebral hypoxia-ischemia in the immature rat. *J Cereb Blood Flow Metab.* 2004;24:1090–1097.
37. Broughton BR, Reutens DC, Sobey CG. Apoptotic mechanisms after cerebral ischemia. *Stroke.* 2009;40:e331–e339.
38. Marcelino TB, de Lemos Rodrigues PI, Miguel PM, Netto CA, Pereira Silva LO, Matté C. Effect of maternal exercise on biochemical parameters in rats submitted to neonatal hypoxia-ischemia. *Brain Res.* 2015;1622:91–101.
39. Ancarcrona M, Dypbuct JM, Bonfoco F, Zhivotovsky B, Orrenius S, Lipton SA, Nicoter P. Glutamat induced neuronal death: a succession of necrosis or apoptosis dependong on mitochondrial function. *Neuron.* 1995;15:961–973.

Chapter 2

Pathophysiology of Asphyxia in Pregnancy

Stefan Dugalic

2.1 Introduction

Pregnancy is a condition in which the body constantly changes. Those changes are normal and expected and occur on both anatomical and physiological levels, affecting every organ and system in the pregnant individual's body (1). During a healthy pregnancy, these changes are reversible and disappear after delivery. Hence, it is very important to know what is happening during pregnancy in order to distinguish the normal, reversible changes from pathological changes.

One of the most important physiological changes is the overall and progressive increase in plasma volume (2). The plasma volume increase is greater than red blood cell expansion, leading to decreased hemoglobin concentration, hematocrit, and erythrocyte count. This condition is known as hemodilution and usually does not change the mean corpuscular volume or mean corpuscular hemoglobin concentration.

DOI: 10.1201/9781003346692-2

The second most common change is the coagulation system, in terms of physiological hypercoagulability, which is a mechanism made to provide hemostasis following delivery (3). Certain clotting factors concentration is increased, particularly factor VIII, IX, and X. Fibrinogen levels increase as well, up to 50%, while fibrinolytic activity is decreased because of the decrease of antithrombin and protein S. Thus, women in postpartum are more prone to vein thrombosis, and this risk lasts up to 12 weeks after delivery. Despite increased clotting activity, prothrombin time (PT) and thrombin time (TT) remain within normal ranges if there is no coagulopathy.

Of course, we have to mention that iron levels change during pregnancy, since the requirements for iron increase two- to threefold, while the folate requirements are even higher (2).

Changes in the cardiovascular system during pregnancy are huge and start from the beginning of pregnancy, somewhere from the eighth week of gestation, with increased cardiac output. Endothelium-dependent factors like nitric oxide and prostaglandins are responsible for peripheral vasodilatation, which decreases vascular resistance by 25–30% and, as a result, increases cardiac output by up to 40% during pregnancy. A further increase is observed during delivery (up to 15% in the first stage and as high as 50% in the second stage) (4).

Pregnancy is associated with increased renal blood flow. Relaxin produced by corpus luteum, decidua, and placenta is elevated, even more after conception, stimulating endothelin formation, which is responsible for vasodilatation of renal arteries induced by nitric oxide (5). Glomerular filtration rate is higher (GFR) by 50–85%, as well as renal plasma flow (40–65%) (6). Renal size is also increasing by 1–1.5 cm, while the kidney, pelvis, and calyceal systems are dilated, and all these changes are mediated by progesterone (6). Dilatation of ureters, renal pelvis, and calyces is present, leading to physiological hydronephrosis in around 80% of women (7). Another common occurrence is the predisposition for urinary infections and pyelonephritis, due to urinary stasis (7). Also, about 90% of women with normal glycoregulation excrete 1–10 g of glucose daily during pregnancy (6).

And of course, we have to mention glucose metabolism changes during pregnancy. Glucose is mandatory for fetal development, but the metabolism also has to work well enough to provide maternal nutrition (8). In early pregnancy, insulin sensitivity is increased and then followed by insulin resistance later (9).

2.2 Embryology and Placenta

Placenta is the fetal organ consisting of parenchyma, chorion, amnion, and umbilical cord. Fetal tissues form from the chorionic sac containing from the amnion, chorion, yolk sac, and allantois. The maternal part is formed from the endometrium and is called decidua, consisting of three parts: decidua basalis, which is implanted deep at the site; decidua capsularis, which covers the site of implantation; and decidua parietalis, comprising everything else (10). Chorion forms the placenta and consists of syncyciotrophoblast, cytotrophoblast, and extraembryonic mesoderm. Cytotrophoblast grows and forms finger-like formations, the so-called primary chorionic villi. Extraembryonic mesoderm splits into somatic and splanchnic mesoderm. Somatic mesoderm further grows into primary villi and forms secondary villi. Mesenchyme is responsible for blood cells and vessels growth, as well as formatting of tertiary villi. Villi are further responsible for capillary bed forming, which connect to the embryo heart. Maternal blood that flows through the capillaries distributes oxygen and nutrients to the fetus. Further growth of the villi leads to branching in the villus chorion and forming of the fetal placenta (11).

Fetomaternal junction provides chorion stability. Chorionic villi attached to decidua basalis form the junction between the chorionic sac and endometrium. Blood vessels in endometrium, so-called spiral arteries, go through a cytotrophoblast shell inside the villi and release maternal blood. This is the way to provide oxygen and nutrients for the fetus. Despite this structure, maternal and fetal blood does not mix (12).

The place where mother and fetus exchange gases and nutrients is the placental membrane, while the umbilical cord attaches the fetus to the placenta and contains two umbilical arteries and one umbilical vein (12).

2.3 Uteroplacental Circulation and Its Specificity

During pregnancy, maternal circulation changes a lot in both structure and function (13, 14) as a consequence of sex steroid hormones influence (15). Blood volume is increased in the breast, kidney, and uterus (16). The function of these changes is to provide adequate oxygen and nutrient supply to the fetus. Nitric oxide, responsible for vasodilatation, changes muscular tone of the uterine artery, as well as other mediators like E2β (17, 18).

Fetal circulation consists of four shunts: ductus venosus, foramen ovale, ductus arterious, and placenta. Transition from fetus to neonate is characterized by closing of these shunts, so they are, in normal circumstances, sealed in a few weeks. Each shunt has its own unique function in fetal circulation (19, 20). Blood with the highest concentration of oxygen and nutrients enters the fetus through the umbilical vein directly to the liver. The first shunt responsible for adequate distribution of nutrients between the liver and central circulation is ductus venosus. The left portal vein is associated with ductus venosus, and that is the spot where blood from the umbilical vein comes in contact with portal blood that drains splanchnic circulation (21). The second organ that receives blood with nutrients is the heart. Circulation through the fetal heart is specific. Ductus venosus and left hepatic vein carry blood with higher concentration of nutrients to the atrial tributaries, compared to other venous tributaries (vena cava superior and anterior, right and middle hepatic veins, and coronary sinus). On the left side, pulmonary veins return depleted blood to the left atrium. Here, we encounter second important shunt – foramen ovale. The position of the foramen ovale valve provides more saturated blood to reach mostly the left ventricle, while less saturated blood goes to the right parts of the heart. The umbilical artery is also a shunt where depleted blood is directed to the placenta to be provided with oxygen, nutrients, and for fluid exchange. The umbilical artery is very important, as it is the only mean for exchange and getting supplies from placenta. Lower resistance in the umbilical artery favors channeling the blood to the placenta. If the resistance in placental blood flow increases, the effect will be the opposite (21). This high-flow–low-resistance circulation on both sides of the placenta (maternal and fetal) is the result of increased cross-section of the vessels due to villous angiogenesis on the fetal and remodeling of spiral arteries and vasodilatation on maternal side (22–25). Longitudinal studies have shown progressive reduction of impedance in uterine artery (26) and in umbilical arteries (27).

2.4 Hypoxia and Hypoxemia in Fetus

Oxygen is necessary for an electron transport chain (ETC). When oxygen level is low, ETC is compromised (28). When oxygen is scarce and its tension is lower than physiological levels, and demand for oxygen exceeds available supply, this condition is called hypoxia. Fetal arterial blood pO_2 is between 20–30 mmHg (29, 30), which is similar to hypoxia in adult tissues, implying that the fetus develops in the condition of relative hypoxia and that fetal tissues have lower threshold for oxygen insufficiency. Given the fact that placenta represents the major organ for oxygen supply, its development and oxygenation have a great impact on fetal development in terms that maternal pO_2 or metabolism changes may reduce fetal arterial pO_2 and induce hypoxia (31, 32). If it enters hypoxia, the fetus is suffering and fights against it in such manner that includes altering fetal blood flow from peripheral tissues to vital organs and a shift from aerobic to anaerobic metabolism.

One of the most common challenges for the fetus, especially during late gestation, is an episode of O_2 deprivation. Reduction of fetal arterial pO_2 of around 50% is also known as acute hypoxemia. This condition induces an integrated cardiovascular, metabolic, and endocrine response aimed at fetal survival during O_2 deprivation (33, 34). During late gestation, cardiovascular response to hypoxemia is triggered by carotid chemoreflex, which leads to bradycardia and redistribution of blood toward vital organs, specifically fetal brain, heart, and adrenal glands (33, 35). Acute fetal hypoxemia is commonly followed by fetal acidemia (36). Hence, there is a concern about a combined effect of hypoxemia and acidemia, compared to the isolated effect of hypoxia (37–40).

2.4.1 Effects of Hypoxia and Hypoxemia, as Well as Impedance Increase, on Fetal Development

Chronic exposure to hypoxia and hypoxemia has a large impact on fetal development, and even leads to health issues later in life. For example, chronic exposure to moderate hypoxia can lead to programming of cardioprotective genes, which may decrease the heart's ability to adapt to stress later in life. Chronic exposure to severe hypoxia significantly impacts cardiomyocytes and their development, leading to cardiomyopathy (41). Animal studies have shown that in a condition of chronic hypoxia, fetus redistributes blood to vital organs: brain, heart, and adrenal glands (42, 43). Doppler velocimetry showed similar results in human (44, 45). Increased cerebral diastolic blood velocity in the fetus represents reduced cerebral vascular resistance, which is a characteristic reaction in the brain during chronic hypoxia. In the terminal phase, it seems that the signs of "brain sparing" disappear in the middle cerebral artery (MCA), but it is not known if this affects all parts of the brain (46). Each part of the human brain has its own important function, but it is very important to emphasize that even here a certain hierarchy is in place, meaning that when the fetus is suffering due to chronic hypoxia, blood redistribution is occurring within the brain itself. This means that blood flow in the brain shifts to the vital and crucial parts of it while the other parts are suffering the lack of oxygenation. The exact mechanism is not known, but it is very important to understand the flow through MCA using Doppler ultrasound and other brain vessels as part of overall fetal surveillance. There are occasions when Doppler velocimetry in MCA is normal while other parts of brain may be jeopardized (47). This same study has shown that in the condition of chronic hypoxia, blood is redistributed to the frontal lobes rather than the lateral and posterior parts of brain (47). The frontal region supplied with blood by the anterior cerebral artery (ACA) plays an important role in intellectual and emotional behavior, while parts supplied by MCA are responsible for motor control and sensory function. Posterior parts of brain are supplied by the posterior cerebral artery (PCA) (48). Other authors also showed that in pregnancies complicated by severe hypertension, blood flow through ACA and MCA was increased (49, 50). Studies on MCA blood velocimetry showed increased end-diastolic flow when the pregnancy is complicated with intrauterine growth restriction (IUGR), associated with fetal hypoxia (51, 52). But the best indicator of IUGR associated with hypoxia is Doppler velocimetry in the umbilical artery (UA), which thus may be very useful as an addition to MCA Doppler velocimetry. Together, both parameters may be used for fetal surveillance in terms that UA predicting IUGR while MCA Doppler velocimetry indicates that the fetus is in the condition of chronic hypoxia (47).

During acute hypoxemia, umbilical blood flow is maintained or even elevated when hypoxemia becomes severe and triggers the increased perfusion pressure and activity of nitric oxide across the umbilical vascular bed (53). Hypoxemia induces metabolic response, which increases the concentration of glucose and lactates (54, 55). Hyperglycemia is a result of combined effects of reduced glucose consumption and increased production (54, 55), while lactic acidosis represents the result of anaerobic metabolism in the peripheral fetal tissues (56). Pituitary gland and pituitary-adrenal axes may also be induced by acute hypoxemia, further inducing sympathoadrenomedullary axes and stress system (57). Hence, the secretion of catecholamines in fetal circulation is promoted, leading to vasoconstriction and glucogenic response (58–60).

Studies have shown that nephrogenesis may be disturbed and altered due to chronic hypoxia (61). Intrauterine hypoxia has an impact on fetal renin-angiotensin system and hemodynamic in fetal kidney (62). So, in which way does fetal hypoxia affect kidney development and potential poor outcomes? And what may happen as a result of fetal hypoxia in terms of kidney function? First, we have to understand the mechanism of nephrogenesis. It is shown that apoptosis and autophagy play an important role in nephrogenesis (63, 64). Apoptosis is adversely involved in renal development, same as autophagy, known as programmed cell death (65). Studies have shown that fetal hypoxia leads to decreased kidney weight and kidney-to–body weight ratio (66). Insufficient blood and oxygen supply induce BCL-2 protein decrease, leading to hypoxia-induced apoptosis in fetal kidney (67). Some studies identify ACE2 expression in the uteroplacental unit as a leading cause of poor kidney development, induced, again, by hypoxia (68). As previously described, intrauterine growth restriction is characterized by an abnormal vascular resistance and compromised blood flow (69).

During normal pregnancy, blood flow progressively increases in the uteroplacental circulation, including the uterine artery and its branches and the umbilical artery. The main uterine arteries provide ~80% of total uteroplacental flow. As a result of proper remodeling, uteroplacental vessels become more resistant to vasoactive substances compared with non-remodeled vessels (70). So, the hypothesis that the fetus redistributes the blood to major organs during hypoxia is also proven here, whereby we can see that kidney development may be jeopardized because of placental insufficiency and inadequate oxygen supply. Abnormal uterine artery resistance can be influenced by incomplete systemic and local adaptations to pregnancy, including endothelial dysfunction and stiffness of maternal vasculature and placental pathology alterations (71). Umbilical artery ultrasound can be used to identify placental insufficiency complications and fetal growth restriction. Clinically, alterations in umbilical flow velocity are more reliable in detecting early-onset fetal growth restriction before 34 weeks of gestation than later in gestation (72). During normal pregnancy, a progressive increase in flow velocity of the umbilical artery from mid- to late gestation is essential for the expansion of the fetoplacental vasculature in the labyrinth as an adaptation to the growing fetus (73). Reduced umbilical artery flow velocity in ACE2 knockout may suggest an incomplete adaptation of the labyrinth.

As a result of poor kidney development in the fetus, amniotic fluid volume (AVF) is reduced. This is known as oligohydramnios. Oligohydramnios complicates 4.4% of all pregnancies at term (74). Oligohydramnios has correlations with any medical or obstetric condition leading to uteroplacental insufficiency (8% of all gestations). Possible causes include chronic hypertension, vascular disease, thrombophilia, and preeclampsia (75).

The management and prognosis of oligohydramnios vary greatly depending on the underlying etiology, the gestational age at diagnosis, and the severity of the condition. Diagnosis of oligohydramnios during the second trimester is more likely to be associated with fetal or maternal anomalies, whereas diagnosis in the third trimester is more likely to be of unexplained origin. In one study, the etiology of oligohydramnios was unexplained in just 4% of second-trimester gestations, whereas 52% of those diagnosed in the third trimester were idiopathic. Only 10.2% of fetuses diagnosed in the second trimester survived, while the survival rate was 85.3% in those diagnosed in the third trimester. In cases of oligohydramnios diagnosed in the second trimester, pulmonary hypoplasia is the most significant predictor of fetal mortality. The mortality rate of second-trimester oligohydramnios can be as high as 90%, with pulmonary hypoplasia accounting for 87% of those deaths. The most severe pulmonary hypoplasia occurs with oligohydramnios before or during 16 to 24 weeks' gestational age, when the terminal sacs of the fetal lung are developing. Low AFV during the second and early third trimesters also increases the likelihood of limb contractures and birth defects due to compression of fetal parts (76).

2.5 Most Important Maternal Conditions Related to Fetal Hypoxia and Placental Insufficiency

2.5.1 Gestational Diabetes Mellitus (GDM) and Preeclampsia (PE)

GDM causes hyperglycemia and hypoxia in the fetus (77), same as PE (78). Hyperglycemia induces the production of reactive oxygen species (ROS) by the mitochondrial electron chain and generates advanced glycation end products, i.e., proteins modified by non-enzymatic glycation, which are the main underlying causes for endothelial dysfunction. Also, hypoxia and fluctuations in oxygen levels are associated with oxidative stress and increased ROS production and disturb endothelial function (79). The adverse biochemical processes induced by hyperglycemia and hypoxia are, in general, of reversible nature. However, offspring of pregnancies exposed to such conditions in utero, i.e., after pregnancies complicated by maternal GDM or PE, possess an increased risk to develop endothelial dysfunction and cardiovascular diseases later in life (80). This fetal programming is also associated with epigenetic changes. For instance, altered DNA methylation has been observed in fetoplacental endothelial cells after GDM pregnancy (81) and in neonatal endothelial progenitor cells after PE pregnancies (82). Underlying causes are multifaceted, but long-lasting effects of hyperglycemia and hypoxia play a role.

2.5.2 Hypertension and PE

In contrast with the slight decrease in blood pressure (BP) during normal pregnancy, some women may show hypertension in pregnancy (HTN-Preg) in one of four forms: chronic HTN that predates pregnancy, preeclampsia (PE)-eclampsia, chronic HTN with superimposed PE, and nonproteinuric gestational HTN (83). Inadequate invasion of cytotrophoblasts into the decidua leads to reduced uteroplacental perfusion pressure (RUPP) and placental ischemia/hypoxia. Placental ischemia causes the release of biologically active factors such as anti-angiogenic factors, inflammatory cytokines, reactive oxygen species, hypoxia-inducible factors, and angiotensin II receptor autoantibodies. These vasoactive factors could cause systemic vascular endotheliosis and consequent increase in vascular resistance and blood pressure, glomerular endotheliosis causing proteinuria, cerebrovascular endotheliosis causing cerebral edema, seizures and visual disturbances, and hepatic endotheliosis, which may contribute to the manifestations of HELLP syndrome. PE-associated vascular endotheliosis causes a decrease in vasodilator mediators such as nitric oxide, prostacyclin, and endothelium-derived hyperpolarizing factor; an increase in vasoconstrictors such as endothelin-1, angiotensin II, and thromboxane A2; and enhanced mechanisms of vascular smooth muscle contraction such as intracellular $Ca2^{+}$, protein kinase C, and Rho-kinase. Changes in matrix metalloproteinase activity and extracellular matrix cause vascular remodeling and further vasoconstriction (84).

2.5.3 Thrombophilia and Placental Insufficiency

Proteins C and S are vitamin K–dependent plasma proteins that work together as an integral part of the body's natural anticoagulant system. They work through the selective inactivation of Factors Va and VIIIa. Mutations in protein C and S genes are the main causative factors for their deficiencies. In addition, acquired deficiencies can also occur due to multiple mechanisms and/or the use of certain drugs (85).
Hypercoagulability due to deficiency of any antithrombotic factors leads to placental thrombosis, hypoperfusion, fetal growth retardation, fetal death, and subsequent fetal losses. Many and colleagues (86) investigated the contribution of

maternal thrombophilia to the pathologic placental features in women with severe pregnancy complications and concluded that placental abnormalities such as infarcts and fibrinoid necrosis were more common in the placentae of women with pregnancy complications and thrombophilia compared to those with the same complications without thrombophilia. In one of the largest studies to date, Rogers and colleagues prospectively collected and examined 105 placentae from women carrying the factor V Leiden mutation and 225 controls matched for maternal age, ethnicity, and hospital (87). They reported hypervascular villi in placentae from women who were heterozygous for the factor V Leiden mutation. In contrast to the previous studies, this study did not demonstrate an increase in the frequency of placental infarcts, small-for-gestational-age placentae, or fetal thrombotic vasculopathy.

2.5.3.1 Follow-up

What can we do to give our contribution to this fetal condition?

1. We can follow up fetal growth and development. Different genetic tests can be performed during pregnancy that are derived from different origins of early embryonic development. Chorionic villus sampling (CVS) is a prenatal procedure in which a sample of chorionic villi is removed from the placenta for genetic testing. Amniocentesis – transabdominal aspiration of amniotic fluid from the uterine cavity – can also be done (88).
2. Ultrasound is an electronic technology that uses the reflection of pulses of high-frequency sound to produce an image. Ultrasound may be used in a variety of circumstances during pregnancy. It has been assumed that the routine use of ultrasound in early pregnancy will result in the earlier detection of problems and improved management of pregnancy complications when compared

with selective use for specific indications such as after clinical complications (e.g., bleeding in early pregnancy) or where there are concerns about fetal growth and development (89).

3. Cardiotocography (CTG) records changes in the fetal heart rate and their temporal relationship to uterine contractions. The aim is to identify babies who may be short of oxygen (hypoxic) to guide additional assessments of fetal well-being or determine if the baby needs to be delivered by caesarean section or instrumental vaginal birth (90).
4. Prenatal non-stress test, popularly known as NST, is a method used to test fetal well-being before the onset of labor. A prenatal NST functions in overall antepartum surveillance with ultrasound as a part or component of the biophysical profile. The presence of fetal movements and fetal heart rate acceleration is the most critical feature of the non-stress test. It is a noninvasive test used for the surveillance of high-risk pregnancies when the fetus is judged clinically to be at risk for hypoxemia or increased risk of death. Trained and certified nurses, midwives, and physicians should read and interpret the NST. The test readings can be reactive and nonreactive. The NST tests can be initiated at 26–28 weeks. It is reactive from 32 weeks. The NST is one of the most widely used techniques for antepartum fetal evaluation. Antepartum fetal surveillance techniques are used to assess the risk of fetal death in pregnancies complicated by preexisting maternal conditions (e.g., diabetes mellitus) as well as those in which complications have developed (e.g., fetal growth restriction) (91). This is the way we follow up the biophysical profile of the fetus.
5. All of the above-mentioned referring to placenta, vascularization, and specificity of fetal blood supply also have to be monitored and followed if we want to prevent poor pregnancy outcome.

2.6 Conclusion

If we consider the uterus and other components that together make up the so-called conceptus as an environment where the fetus should be safe and have everything needed for its growth and development, it is clear why it is so important to understand the physiology and pathophysiology of placenta – including both the maternal and fetal parts. In conditions when the fetus becomes compromised in any way, including placental insufficiency or poor oxygen and nutrient supply, it will start to compensate this condition and to fight against it. We have already described blood flow shift toward major organs – heart, brain, and adrenal gland – and less blood supply for peripheral tissues. But this mechanism is only temporary and compromises the fetus in many ways. Small-for-gestational age, and oligohydramnios are some of the consequences of the compromised blood supply. When the fetus cannot compensate the conditions in which it finds itself, or if the obstetrician evaluates that the environment becomes hostile, delivery is indicated. Depending on many factors, including the ones mentioned in this chapter, and assessments made by obstetrician, delivery can be done vaginally or surgically (cesarian section).

References

1. Locktich G. Clinical biochemistry of pregnancy. *Crit Rev Clin Lab Sci.* 1997;34:36.
2. Rodger M, Sheppard D, Gandara E, et al. Haematological problems in obstetrics. *Best Prac Res Clin Obstet Gynaecol.* 2015;29(5):671–684.
3. Ramsay M. Normal hematological changes during pregnancy and the puerperium. In: Pavord S, Hunt B, editors. *The Obstetric Hematology Manual.* Cambridge: Cambridge University Press; 2010, pp. 3–12.

4. Soma-Pillay P, Nelson-Piercy C, Tolppanen H, et al. Physiological changes in pregnancy. *Cardiovasc J Afr.* 2016;27(2):89–94.
5. Conrad KP. Emerging role of relaxin in the maternal adaptations to normal pregnancy: implications for preeclampsia. *Semin Nephrol.* 2011;31(1):15–32.
6. Cheung KL, Lafayette RA. Renal physiology of pregnancy. *Adv Chronic Kidney Dis.* 2013;20(3):209–214.
7. Rasmussen PE, Nielsen FR. Hydronephrosis during pregnancy: a literature survey. *Eur J Obstet Gynecol Reprod Biol.* 1988;27(3):249–259.
8. Angueira AR, Ludvik AE, Reddy TE, et al. New insights into gestational glucose metabolism: lessons learned from 21st century approaches. *Diabetes.* 2015;64:327–334.
9. Butte NF. Carbohydrate and lipid metabolism in pregnancy: normal compared with gestational diabetes mellitus. *Am J Clin Nutr.* 2000;71:125.
10. Burton GJ, Jauniaux E. What is the placenta? *Am J Obstet Gynecol.* 2015;213(4):S6.e1, S6–S8.
11. Solnica-Krezel L, Sepich DS. Gastrulation: making and shaping germ layers. *Annu Rev Cell Dev Biol.* 2012;28:687–717.
12. Labarrere CA, DiCarlo HL, Bammerlin E, et al. Failure of physiologic transformation of spiral arteries, endothelial and trophoblast cell activation, and acute atherosis in the basal plate of the placenta. *Am J Obstet Gynecol.* 2017;216(3):287.e1–287.e16.
13. Thornburg K, Bagby S, Giraud G. Maternal adaptation to pregnancy In: Neill JD, editor. *Knobil & Neill's Physiology of Reproduction.* 3rd ed London: Elsevier Adademic Press; 2006. pp. 2899–2924.
14. Thornburg KL, Jacobson SL, Giraud GD, et al. Hemodynamic changes in pregnancy. *Semin Perinatol.* 2000;24:11–14.
15. Hall JM, Couse JF, Korach KS. The multifaceted mechanisms of estradiol and estrogen receptor signaling. *J Biol Chem* 2001;276:36869–36872.
16. Thornburg KL, Louey S. Uteroplacental circulation and fetal vascular function and development. *Curr Vasc Pharmacol.* 2013;11(5):748–757.
17. Rupnow HL, Phernetton TM, Shaw CE, et al. Endothelial vasodilator production by uterine and systemic arteries. VII. Estrogen and progesterone effects on eNOS. *Am J Physiol Heart Circ Physiol.* 2001;280:H1699–H1705.

18. Gibson TC, Phernetton TM, Wiltbank MC, et al. Development and use of an ovarian synchronization model to study the effects of endogenous estrogen and nitric oxide on uterine blood flow during ovarian cycles in sheep. *Biol Reprod.* 2004;70:1886–1894.
19. Schneider DJ. The patent ductus arteriosus in term infants, children, and adults. *Semin Perinatol.* 2012;36:146–153.
20. Tchirikov M, Schroder HJ, Hecher K. Ductus venosus shunting in the fetal venous circulation: regulatory mechanisms, diagnostic methods and medical importance. *Ultrasound Obstet Gynecol.* 2006;27:452–461.
21. Kilavuz O, Vetter K, Kiserud T, Vetter P. The left portal vein is the watershed of the fetal venous system. *J Perinat Med.* 2003;31:184–187.
22. Brosens I, Robertson WB, Dixon HG. The physiologicalresponse of the vessels of the placental bed to normal pregnancy. *J Pathol Bacteriol.* 1967;93:569–579.
23. De Wolf F, Wolf-Peeters C, Brosens I. Ultrastructure of thespiral arteries in the human placental bed at the end of normalpregnancy. *Am J Obstet Gynecol.* 1973;117:833–848.
24. Heppard BL, Bonnar J. The ultrastructure of the arterialsupply of the human placenta in early and late pregnancy. *J Obstet Gynaecol Br Commonw.* 1974;81:497–511.
25. Robertson WB, Brosens I, Dixon G. Utero placental vascular pathology. *Eur J Obstet Gynecol Reprod Biol.* 1975;5:47–65.
26. Gomez O, Figueras F, Fernandez S, et al. Reference ranges for uterineartery mean pulsatility index at 11–41 weeks of gestation. *Ultrasound Obstet Gynecol.* 2008;32:128–132.
27. Acharya G, Wilsgaard T, Berntsen GK, et al. Reference ranges for serial measurements ofumbilical artery Doppler indices in the second half of pregnancy. *Am J Obstet Gynecol.* 2005;192:937–944.
28. Webster WS, Abela D. The effect of hypoxia in development. *Birth Defects Res C Embryo Today.* 2007;81:215–228.
29. Lawrence J, Xiao D, Xue Q, et al. Prenatal nicotine exposure increases heart susceptibility to ischemia/reperfusion injury in adult offspring. *J Pharmacol Exp Ther.* 2008;324:331–341.
30. Mitchell JA, Van Kainen BR. Effects of alcohol on intrauterine oxygen tension in the rat. *Alcohol Clin Exp Res.* 1992;16:308–310.
31. Myatt L. Placental adaptive responses and fetal programming. *J Physiol.* 2006;572:25–30.

32. Jensen A, Garnier Y, Berger R. Dynamics of fetal circulatory responses to hypoxia and asphyxia. *Eur J Obstet Gynecol Reprod Biol.* 1999;84:155–172.
33. Giussani DA, Spencer JA, Hanson MA. Fetal cardiovascular reflex responses to acute hypoxaemia. *Fetal Maternal Med Rev.* 1994;6:17–37.
34. Rudolph AM. Distribution and regulation of blood flow in the fetal and neonatal lamb. *Circ Res.* 1985 57:811–821.
35. Bennet L, Peebles DM, Edwards AD, et al. The cerebral hemodynamic response to asphyxia and hypoxia in the near-term fetal sheep as measured by near infrared spectroscopy. *Pediatr Res.* 1998;44:951–957.
36. Nicolaides KH, Economides DL, Soothill PW. Blood gases, pH, and lactate in appropriate- and small-for-gestational-age fetuses. *Am J Obstet Gynecol.* 1989;161:996–1001.
37. Dweck HS, Huggins W, Dorman LP, et al. Developmental sequelae in infants having suffered severe perinatal asphyxia. *Am J Obstet Gynecol.* 1974;119:811–815.
38. Fee SC, Malee K, Deddish R, et al. Severe acidosis and subsequent neurologic status. *Am J Obstet Gynecol.* 1990;162:802–806.
39. Ruth VJ, Raivio KO. Perinatal brain damage: predictive value of metabolic acidosis and the Apgar score. *Br Med J.* 1988;297:24–27.
40. Soothill PW, Ajayi RA, Campbell S, et al. Fetal oxygenation at cordocentesis, maternal smoking and childhood neurodevelopment. *Eur J Obstet Gynecol Reprod Biol.* 1995;59:21–24.
41. Patterson AJ, Zhang L. Hypoxia and fetal heart development. *Curr Mol Med.* 2010;10(7):653–666.
42. Cohn HE, Sacks EJ, Heymann MA, et al Cardiovascularresponses to hypoxemia and acidemia in fetal lambs. *Am J ObstetGynecol.* 1974;120:817–824.
43. Sheldon RE, Peeters LL, Jones MD Jr, et al. Redistribution of cardiac output and oxygen delivery in the hypoxemicfetal lamb. *Am J Obstet Gynecol.* 1979;135:1071–1077.
44. Mari G, Uerpairojkit B, Abuhamad AZ, et al. Adrenal artery velocity waveforms in the appropriate and small-for-gestational age fetus. *Ultrasound Obstet Gynecol.* 1996;8:82–86.
45. Gembruch U, Baschat AA. Demonstration of fetal coronary blood flow by color-coded and pulsed wave doppler sonography: a possible indicator of severe compromise and impending

demise in intra uterine growth retardation. *Ultrasound Obstet Gynecol.* 1996;7:10–15.
46. Weiner Z, Farmakides G, Schulman H, et al. Central and peripheral hemodynamic changes in fetuses withabsent end-diastolic velocity in umbilical artery: correlation withcomputerized fetal heart rate pattern. *Am J Obstet Gynecol.* 1994;170:509–515.
47. Dubiel M, Gunnarsson GO, Gudmundsson S. Blood redistribution in the fetal brain during chronic hypoxia. *Ultrasound Obstet Gynecol.* 2002;20(2):117–121.
48. Okudera T, Ohta T, Huang YP, Yokota A. Developmental and radiological anatomy of the superficial cerebral convexity vessels in the human fetus. *J Neuroradiol.* 1988;15:205–224.
49. Arbeille P, Body G, Saliba E. Fetal cerebral circulation assessment by Doppler ultrasound in normal and pathological pregnancies. *Eur JObstet Gynecol Reprod Biol.* 1988;29:261–273.
50. Noordam MJ, Heydanus R, Hop WC, et al. Doppler colour flow imaging of fetal intracerebral arteries and umbilical artery in the small for gestational age fetus. *Br J Obstet Gynaecol.* 1994;101:504–508.
51. Woo JS, Liang ST, Lo RL, et al. Middle cerebral artery Doppler flow velocity waveforms. *Obstet Gynecol.* 1987;70:613–616.
52. Vyas S, Nicolaides KH, Bower S, et al. Middle cerebral artery flow velocity waveforms in fetal hypoxaemia. *Br J Obstet Gynaecol.* 1990;97:797–803.
53. Thakor AS, Giussani DA. Calcitonin gene-related peptide contributes to the umbilical haemodynamic defence response to acute hypoxaemia. *J Physiol.* 2005;563:309–317.
54. Jones CT. The development of some metabolic responses to hypoxia in the foetal sheep. *J Physiol.* 1977; 265:743–762.
55. Jones CT, Ritchie JW, Walker D. The effects of hypoxia on glucose turnover in the fetal sheep. *J Dev Physiol.* 1983;5:223–235.
56. Boyle DW, Hirst K, Zerbe GO, et al. Fetal hind limb oxygen consumption and blood flow during acute graded hypoxia. *Pediatr Res.* 1990;28:94–100.
57. Chrousos GP. Stressors, stress, and neuroendocrine integration of the adaptive response. *Ann NY Acad Sci.* 1998;851:311–335.
58. Slotkin TA, Seidler FJ. Adrenomedullary catecholamine release in the fetus and newborn: secretory mechanisms and their role in stress and survival. *J Dev Physiol.* 1988;10:1–16.

59. Jones CT, Wei G. Adrenal-medullary activity and cardiovascular control in the fetal sheep. In: Kunzel W, editor. *Fetal Heart Rate Monitoring*. Berlin: Springer-Verlag; 1985, pp. 127–135.
60. Apatu RS, Barnes RJ. Release of glucose from the liver of fetal and postnatal sheep by portal vein infusion of catecholamines or glucagon. *J Physiol.* 1991;436:449–468.
61. Gonzalez-Rodriguez PJ, Tong W, Xue Q, et al. Fetal hypoxia results in programming of aberrant angiotensin II receptor expression patterns and kidney development. *Int J Med Sci.* 2013;10(5):532–538.
62. Allam HI, Rahman TA. The protective role of adrenomedullin and anti-tumor necrosis factor on some vasoregulatory factors and renal functions in hypertensive pregnant rats. Kingdom of Saudi Arabia King Abdulaziz City for Science and Technology General Directorate of Research Grants Programs LGP-10-19 final report. 2009:43–78.
63. Koseki C, Herzlinger D, Al-Awqati Q. Apoptosis in metanephric development. *J Cell Biol.* 1992;119(5):1327–1333.
64. Coles HS, Burne JF, Raff MC. Large-scale normal cell death in the developing rat kidney and its reduction by epidermal growth factor. *Development.* 1993;118(3):777–784.
65. Sorenson CM. Life, death and kidneys: regulation of renal programmed cell death. *Curr Opin Nephrol Hypertens.* 1998;7(1):5–12.
66. Xia S, Lv J, Gao Q, et al. Prenatal exposure to hypoxia induced Beclin 1 signaling-mediated renal autophagy and altered renal development in rat fetuses. *Reprod Sci.* 2015;22(2):156–164.
67. Lee P, Sata M, Lefer DJ, et al. FAS pathway is a critical mediator of cardiac myocyte death and myocardial infarction during ischemia–reperfusion in vivo. *Am J Physiol Heart Circ Physiol.* 2003;284(2):H456–H463.
68. Neves LAA, Stovall K, Joyner J, et al. ACE2 and Ang-(1–7) in the uterus during early and late gestation. *Am J Physiol Regul Integr Comp Physiol.* 2008;294:151–161.
69. Khalil RA, Granger JP. Vascular mechanisms of increased arterial pressure in preeclampsia: lessons from animal models. *Am J Physiol Regul Integr Comp Physiol.* 2002; 283:29–45.
70. Osol G, Mandala M. Maternal uterine vascular remodeling during pregnancy. *Physiology (Bethesda).* 2009;24:58–71.

71. Everett TR, Lees CC. Beyond the placental bed: placental and systemic determinants of the uterine artery Doppler waveform. *Placenta*. 2012;33:893–901.
72. Mayer C, Joseph KS. Fetal growth: a review of terms, concepts and issues relevant to obstetrics. *Ultrasound Obstet Gynecol*. 2013;41:136–145.
73. Croy BA, Yamada AT, DeMayo FJ, Adamson SL, eds. *The Guide to Investigation of Mouse Pregnancy*. 1st ed. Amsterdam: Elsevier, 2014.
74. Hou L, Wang X, Hellerstein S, et al. Delivery mode and perinatal outcomes after diagnosis of oligohydramnios at term in China. *J Matern Fetal Neonatal Med*. 2020;33(14):2408–2414.
75. Peipert JF, Donnenfeld AE. Oligohydramnios: a review. *Obstet Gynecol Surv*. 1991;46(6):325–339.
76. Shipp TD, Bromley B, Pauker S, et al. Outcome of singleton pregnancies with severe oligohydramnios in the second and third trimesters. *Ultrasound Obstet Gynecol*. 1996;7(2):108–113.
77. Escobar J, Teramo K, Stefanovic V, et al. Amniotic fluid oxidative and nitrosative stress biomarkers correlate with fetal chronic hypoxia in diabetic pregnancies. *Neonatology*. 2013;103:193–198.
78. Teramo KA, Hiilesmaa VK, Schwartz R, et al. Amniotic fluid and cord plasma erythropoietin levels in pregnancies complicated by preeclampsia, pregnancy-induced hypertension and chronic hypertension. *J Perinat Med*. 2004;32:240–247.
79. Giacco F, Brownlee M. Oxidative stress and diabetic complications. *Circ Res*. 2010;107:1058–1070.
80. Frost AL, Suriano K, Aye CYL, et al. The immediate and long-term impact of preeclampsia on offspring vascular and cardiac physiology in the preterm infant. *Front Pediatr*. 2021;9:625726.
81. Cvitic S, Novakovic B, Gordon L, et al. Human fetoplacental arterial and venous endothelial cells are differentially programmed by gestational diabetes mellitus, resulting in cell-specific barrier function changes. *Diabetologia*. 2018;61:2398–2411.
82. Brodowski L, Zindler T, von Hardenberg S, et al. Preeclampsia-associated alteration of DNA methylation in fetal endothelial progenitor cells. *Front Cell Dev Biol*. 2019;7:32.
83. Shah DA, Khalil RA. Bioactive factors in uteroplacental and systemic circulation link placental ischemia to generalized

vascular dysfunction in hypertensive pregnancy and pre-eclampsia. *Biochem Pharmacol.* 2015;95:211–226.
84. Ali SM, Khalil RA. Genetic, immune and vasoactive factors in the vascular dysfunction associated with hypertension in pregnancy. *Expert Opin Ther Targets.* 2015;19(11):1495–1515.
85. Esmon C, Vigano D, Angelo S, et al. Anticoagulation proteins C and S. *Advanced Exp Med Biol.* 1987;214:47–54.
86. Manu A, Schreiber L, Rosner S et al. Pathologic features of the placenta in women with severe pregnancy complications and thrombophilia. *Obstet Gynecol.* 2001;98(6):1041–1044.
87. Rogers BB, Momirova V, Dizon-Townson D, et al. Avascular villi, increased syncytial knots, and hypervascular villi are associated with pregnancies complicated by factor V Leiden mutation. *Pediatr Dev Pathol.* 2010;13(5):341–347.
88. Ghi TA, Sotiriadis P, Calda F, et al. International society of ultrasound in and gynecology ISUOG: invasive procedures for prenatal diagnosis. *Ultrasound Obstet Gynecol.* 2016;48:256–268.
89. Whitworth M, Bricker L, Mullan C. Ultrasound for fetal assessment in early pregnancy. *Cochrane Database Syst Rev.* 2015;2015(7):CD007058.
90. Alfirevic Z, Devane D, Gyte GM, Cuthbert A. Continuous cardiotocography (CTG) as a form of electronic fetal monitoring (EFM) for fetal assessment during labour. *Cochrane Database Syst Rev.* 2017;2(2):CD006066.
91. Preboth M. ACOG guidelines on antepartum fetal surveillance. American College of Obstetricians and Gynecologists. *Am Fam Physician.* 2000;62(5):1184, 1187–1188.

Chapter 3

Causes of Perinatal Asphyxia

Sandra Babic

3.1 Introduction

Perinatal asphyxia is defined as a gas exchange disorder at the level of the fetomaternal unit, in the placenta itself, in the fetus during birth, or in the lungs of the newborn after birth. Fetal hypoxia can occur during pregnancy or during childbirth. The appearance of fetal hypoxia in the early period of pregnancy causes damage and slows down the development of the embryo. A number of conditions can predispose a baby to birth asphyxia, but the underlying etiology is reduced cerebral blood flow (1).

Factors that lead to asphyxia are maternal conditions (bleeding, amniotic fluid embolism or high blood pressure, diabetes mellitus, and other maternal diseases), factors at the level of the placenta (acute abruption), in the uterus (rupture), events in the umbilical cord (long or short umbilical

DOI: 10.1201/9781003346692-3

cord or umbilical cord prolapse), intrapartum infection, and very long or delayed labor (2, 3).

There is a direct correlation with maternal bleeding before delivery, in the sense of abruption of moderate or severe first-degree asphyxia at birth. An explanation for the above could be that bleeding during pregnancy reduces uteroplacental perfusion, which leads to asphyxia at birth (4, 5).

Premature rupture of fetal membranes is also positively associated with asphyxia at birth because due to premature rupture of membranes, the umbilical cord is no longer protected by the amniotic fluid and can become compressed. Compression of the umbilical cord partially or completely cuts off the flow of oxygen-rich blood to the baby, who asphyxiates at birth (6).

Another justification may be that without protection of the fetus, infections of the mother can be transmitted to the baby, leading to sepsis of the newborn followed by birth asphyxia (7).

It was found that babies with low birth weight have a higher risk of developing asphyxia compared to babies with normal birth weight, which results in the intrauterine release of meconium into the amniotic fluid. The presence of meconium in the amniotic fluid increases the risk of aspiration of meconium during intrauterine breathing or the first breaths after birth, which can cause acute airway obstruction, dysfunction, or inactivation of surfactant. Meconium in the amniotic fluid is found in 4–22% of all births, and in 23–52% of deliveries after 42 weeks of gestation. It is notable that 3–12% of babies who have small amounts of meconium in the amniotic fluid develop meconium aspiration syndrome MAS (6).

Asphyxia occurs more often in women who have prolonged labor and in multiparous women than in women who give birth for the first time. The most likely reason for this is the fact that women who have already had prolonged labor have more negative birth experiences than women who have undergone normal labor. Neonates born from prolonged labor experience stress due to a large number of uterine contractions as well as problems of pressure on the umbilical cord (8).

Fetal presentation increases the risk of a number of complications, including umbilical cord prolapse or compression, which can lead to birth asphyxia (9). Some maternal conditions, such as anemia, have been found to be a significant risk factor for birth asphyxia. Lower maternal hemoglobin concentrations during labor are associated with poor fetal oxygen perfusion and a higher risk of asphyxia during labor (10). Premature babies are more likely to develop birth asphyxia than are full-term babies. The reason for this could be that the organs of a premature baby are not yet fully mature, which means that the respiratory system, especially the baby's lungs, cannot work optimally due to an inadequate amount of surfactant (11).

Birth asphyxia is associated with cesarean delivery. A possible reason for this finding is the fact that during vaginal delivery, compression of the baby's chest removes approximately one-third of fetal lung fluid, which does not occur during cesarean delivery. In addition, less surfactant is secreted to the alveolar surface during cesarean delivery, leading to a higher risk of developing asphyxia during delivery (12).

The probability of developing birth asphyxia is 2.52 times higher in women who have chronic hypertension than in women who do not. This finding is almost universal, with pregnancies complicated by chronic hypertension reported to result in adverse outcomes such as low birth weight, preterm birth, perinatal death, and birth asphyxia (13).

Delivery after term was significantly associated with an increased likelihood of birth asphyxia, with complications for the newborn, and aspiration of meconium, which can lead to intrauterine asphyxia and asphyxia during childbirth (14).

Women who underwent induction were more likely to develop asphyxia at birth than those who did not. A possible reason for this finding may be due to the fact that induction drugs can cause excessive uterine contractions (hyperstimulation), which can lead to changes in the fetal heart rate. When contractions are too fast and strong, the placenta, which helps carry oxygen-rich blood to the baby, often cannot be replenished with an adequate supply of this blood for the baby (15).

Prolonged labor in the second stage of labor is associated with an increased risk of birth asphyxia. The second stage of labor is characterized by high contraction of the uterus and implies a period of relative lack of oxygen, since the baby's head and umbilical cord are compressed by contractions in the birth canal, so that in the prolonged second stage of labor, the condition of insufficient oxygenation worsens and leads to asphyxia during childbirth (16).

In late pregnancy, hypoxia slows down growth, affects the central nervous system, and slows down the adaptive abilities of the newborn. Fetal hypoxia can occur due to harmful processes taking place in the body of the child, of the mother, or in the placenta. Sometimes during delivery, the umbilical cord can wrap around the baby's neck, causing breathing to stop. If a baby stops breathing for just a minute, they may have hypoxia, while a baby who stops breathing for close to 5 minutes is likely to have anoxia. A less common situation during childbirth is umbilical cord prolapse when the umbilical cord comes out of the cervix ahead of the baby. If the umbilical cord is pressed or twisted, the baby may be deprived of oxygen supply, and in this case doctors must urgently facilitate the delivery via cesarean section (17). Fetal hypoxia can occur as a result of impaired oxygen transport, before delivery, during delivery, and after delivery. Causes are divided into those of maternal origin, placental origin, during childbirth, fetal, and neonatal (18).

3.2 Causes of Perinatal Asphyxia (Hypoxia)

3.2.1 Hypoxia of Maternal Origin

Maternal hypoxia occurs as a result of insufficient oxygenation or a decrease in maternal blood volume, disturbances in the metabolism of nutrients and mineral substances in chronic respiratory insufficiency, chronic decompensated cardiomyopathy, shock, essential arterial hypertension, chronic renal insufficiency, arterial hypotension, chronic alcoholism, poor socioeconomic conditions of life, and complications due to anesthesia (19).

3.2.2 Hypoxia of Placental Origin

Hypoxia in placental disorders occurs due to insufficient function of the placenta (placental insufficiency), in the case of biologically transmitted pregnancy, diabetes mellitus, primipara vetusta (older pregnant woman), placenta previa, premature detachment of the normally implanted placenta, placental anomalies, and intrauterine infections (toxoplasmosis, listeriosis, viral and bacterial infections) (20).

3.2.3 Hypoxia during Childbirth

It occurs in the following situations: abruption of the placenta, placenta previa, prolapse of the umbilical cord, pressure of the fetus on the umbilical cord, the umbilical cord wrapped around the neck or body of the fetus, irregular position of the fetus during childbirth, pelvic presentation of the fetus, legged position of the fetus, surgical termination (cesarean section), irregular uterine contractions, use of instrumental interventions during childbirth (forceps), and, rarely, sudden death (21).

3.2.4 Hypoxia of Fetal Origin

Hypoxia of fetal origin occurs due to insufficient oxygenation or insufficient oxygen transport, fetal anemia, increased oxygen consumption in complications with the umbilical cord, fetal anomalies, anemia due to hemolytic disease, prematurity, ascending intrauterine infections, and respiratory depression due to anesthesia or analgesia of the mother (22).

3.2.5 Hypoxia of the Newborn

The most common causes of newborn hypoxia are: congenital anomalies of the respiratory tract; congenital diaphragmatic hernia; intrauterine pneumonia; intrapartum aspiration of amniotic fluid; damage to the brain, medulla oblongata, or phrenic nerve; and depression of the brain stem by analgesics and spasmolytics given to the mother during childbirth (23).

Although listed like this (for didactic reasons), very often newborn asphyxia is followed by asphyxia during childbirth. Acute fetal asphyxia is rare and comes down to an obstetric emergency such as umbilical cord prolapse or placental abruption. Most infants show signs of a subacute or chronic process of intermittent hypoxic episodes (24). These signs can be recognized by ultrasound examination of a pregnant woman, where oligohydramnios or intrauterine growth retardation is an important indicator of fetal prenatal asphyxia/hypoxia. To assess the risk of asphyxia, a fetal biophysical profile has been established, whereby five ultrasound-determined biophysical variables are monitored (fetal breathing movements, limb movements, tone, heart rate, and determination of the amount of amniotic fluid) (25).

An important factor in the assessment of asphyxia is the equipment and training of obstetric and neonatology resuscitation teams in operating and delivery rooms, which can directly shorten or prolong fetal or neonatal hypoxia. In recent years in our country (Serbia), and in recent decades in developed countries, with a well-trained perinatal team (obstetrician, neonatologist, nurse), the frequency of asphyxia has decreased. However, there remains a small percentage of unpredictable cases that are difficult to prevent even with the use of state-of-the-art equipment and resuscitation procedures (26).

3.3 Conclusion

The most important factor in the occurrence of asphyxia, as well as the etiological cause, is the basis of caring for a newborn affected by asphyxia. The consequences, as well as the treatment, are positively correlated with the knowledge of the cause of asphyxia.

References

1. Roberto A, Annalisa P, Maria DP. Perinatal asphyxia in the term newborn. *J Pediatric Neonatal Individ Med.* 2014;3(2):e030269 (Ch3).
2. Tegegnework SS, Gebre YT, Ahmed SM, Tewachew AS. Determinants of birth asphyxia among newborns in Debre Berhan referral hospital, Debre Berhan, Ethiopia: a case-control study. *BMC Pediatr.* 2022;22:165.
3. Ahmed R, Mosa H, Sultan M, Helill SE, Assefa B, Abdu M, Ahmed U, Abose S, Nuramo A, Alemu A, Demelash M, Delil R. Prevalence and risk factors associated with birth asphyxia among neonates delivered in Ethiopia: a systematic review and meta-analysis. *PLoS One.* 2021;16(8):e0255488.
4. Linde LE, Rasmussen S, Kessler J, Ebbing C. Extreme umbilical cord lengths, cord knot and entanglement: Risk factors and risk of adverse outcomes, a population-based study. *PLoS One.* 2018;13:e0194814.
5. Wu YW, Colford JM Jr: Chorioamnionitis as a risk factor for cerebral palsy: a meta-analysis. *JAMA.* 2000;284:1417.

6. Monfredini C, Cavallin F, Villani PE, Paterlini G, Allais B, Trevisanuto D. Meconium aspiration syndrome: a narrative review. *Children (Basel)*. 2021;8:230.
7. Fischer CP, Rybakowski C, Ferdynus C, Sagot P, Gouyon JB. A population-based study of meconium aspiration syndrome in neonates born between 37 and 43 weeks of gestation. *Int J Pediatr*. 2012;2012:321545.
8. Nystedt A, Hildingsson I. Diverse definitions of prolonged labour and its consequences with sometimes subsequent inappropriate treatment. *BMC Pregnancy and Childbirth*. 2014;14:233.
9. Yadav N, Damke S. Study of risk factors in children with birth asphyxia. *Int J Contemp Pediatr*. 2017;4:518–526.
10. Pradhan P, Mohajer M, Deshpande S. Outcome of term breech births: 10-year experience at a district general hospital. *BJOG*. 2005;112:218–222.
11. Akella A, Deshpande SB. Pulmonary and their role in pathophysiology of lung disorders. *Indian J Exp Biol*. 2013;51:5–22.
12. Maria Cecilia Santa Cruz Breim, Conceição Aparecida de Mattos Segre, Umberto Gazi Lippi. Morbidity in neonates according to the mode of delivery: a comparative study. *Einstein*. 2010;8:308–314.
13. Agrawal A, Wenger NK. Hypertension During Pregnancy. *Curr Hypertens Rep*. 2020;22:64.
14. Berhe YZ, Kebedom AG, Gebregziabher L, Assefa NE, Berhe LZ, Mohammednur SA, Wellay T, Berihu G, Welearegay AT, Mitiku M, Teka HG. Risk factors of birth asphyxia among neonates born in public hospitals of Tigray, Northern Ethiopia. *Pediatric Health Med Ther*. 2020;11:13–20.
15. Rydahl E, Eriksen L, Juhl M. Effects of induction of labor prior to post-term in low-risk pregnancies: a systematic review. *JBI Database System Rev Implement Rep*. 2019;17:170–208.
16. Vayssière C, Haumonte JB, Chantry A, Coatleven F, Debord MP, Gomez C, Le Ray C, Lopez E, Salomon LJ, Senat MV, Sentilhes L, Serry A, Winer N, Grandjean H, Verspyck E, Subtil D, French College of Gynecologists and Obstetricians (CNGOF). Prolonged and post-term pregnancies: guidelines for clinical practice from the French College of Gynecologists and Obstetricians (CNGOF). *Eur J Obstet Gynecol Reprod Biol*. 2013;169:10–16.

17. Fahey J, King TL. Intrauterine asphyxia: clinical implications for providers of intrapartum care.*J Midwifery Womens Health*. 2005;50:498–506.
18. Mota-Rojas D, Villanueva-García D, Solimano A, Muns R, Ibarra-Ríos D, Mota-Reyes A. Pathophysiology of perinatal asphyxia in humans and animal models. *Biomedicines*. 2022;10:347.
19. Marcelino TB, de Lemos Rodrigues PI, Miguel PM, Netto CA, Pereira Silva LO, Matté C. Effect of maternal exercise on biochemical parameters in rats submitted to neonatal hypoxia-ischemia. *Brain Res*. 2015;1622:91–101.
20. Sakornbut E, Leeman L, Fontaine P. Late pregnancy bleeding. *Am Fam Physician*. 2007;75:1199–1206.
21. Bruckmann EK, Velaphi S. Intrapartum asphyxia and hypoxic ischaemic encephalopathy in a public hospital: Incidence and predictors of poor outcome. *S Afr Med J*. 2015;105:298–303.
22. Torres-Muñoz J, Rojas C, Mendoza-Urbano D, Marín-Cuero D, Orobio S, Echandía C. Risk factors associated with the development of perinatal asphyxia in neonates at the Hospital Universitario del Valle, Cali, Colombia, 2010–2011. *Biomedica*. 2017;37:51–56.
23. Low JA. Determining the contribution of asphyxia to brain damage in the neonate. *J Obstet Gynaecol Res*. 2004;30:276–286.
24. Manning FA. Antepartum fetal testing: a critical appraisal.*Curr Opin Obstet Gynecol*. 2009;21:348–352.
25. Manning FA. Fetalbiophysicalprofile: a critical appraisal. *Clin Obstet Gynecol*. 2002;45:975–985.
26. Hankins GD, Koen S, Gei AF, Lopez SM, Van Hook JW, Anderson GD. Neonatal organ system injury in acute birth asphyxia sufficient to result in neonatal encephalopathy.*Obstet Gynecol*. 2002;99:688–691.

Chapter 4

Multiorgan Postasphyctic Damage

Miljana Z. Jovandaric

4.1 Introduction

In conditions of hypoxia and ischemia due to reduced production of adenosine triphosphate (ATP), many structural and functional changes occur, especially in highly differentiated cells of the central nervous system (CNS), heart, and kidneys. It is certain that the loss of cell membrane function is the primary event in the occurrence of cell damage under hypoxic conditions. Numerous experimental studies support the importance of free radical production after acute hypoxia. All metabolic changes associated with reversible changes in cell hypoxia, if qualitatively or quantitatively significant, can cause irreversible damage (1).

In hypoxia, the tissue that was in normal conditions in a dynamic equilibrium, on the one hand with contact with substances and on the other with removal of waste products, now turns into a closed system. Oxidative phosphorylation

 DOI: 10.1201/9781003346692-4

and metabolism of oxygen dependent substances stop, metabolic changes occur depending on internal reserves that are quickly consumed, and metabolic end products harmful to the tissue accumulate, which leads to the destruction of cells and tissues (2).

In hypoxia, compared to the normal state of oxygen exposure, the life conditions of the fetus change. Acute or chronic suffering of the fetus occurs. Fetal hypoxia can occur as a result of disturbed oxygen transport at various stages: maternal hypoxia, reduction of uterine blood flow, disruption of placental function, and changes in the umbilical cord. In the placenta, the microcirculation is disturbed, which causes a metabolic disorder in the mother–fetus system, because there is a disturbance in the exchange of nutrients and, later, in the transport of oxygen. Fetal hypoxia occupies an important place in the causes of postnatal mortality, as well as somatic and neurological disorders of children. Hypoxia is the primary cause of various consequent pathological conditions. The essence of the hypoxic process is energy deficit, which leads to functional and morphological changes and ultimately to cell death (3).

4.1.1 Immediate Effects of Asphyxia

Hypoxia disrupts cellular metabolism in many different ways. The main role is played by the limited reserve of glycogen. Under aerobic conditions, glycolysis (through the Krebs cycle and oxidative phosphorylation) produces 38 molecules of ATP from 1 molecule of glucose. ATP powers the neuronal membrane pump that keeps the internal concentration of potassium high and sodium, calcium, and chloride low. This ionic balance is the physiological basis for intraneuronal communication. With the lack of glucose and oxygen, all ATP-dependent processes are interrupted. In complete hypoxia/ischemia, the cell consumes all glycogen and all ATP stores within 10 minutes. Anaerobic metabolism begins in the cell (it produces only two molecules of ATP from each molecule of glucose), which leads to a decrease in pH. Without ATP, the ion pump loses function, sodium and chlorine enter the cell, and then water follows by passive diffusion with the ultimate effect of cytotoxic cellular edema. The cell swells, ruptures, and releases its cytoplasmic contents into the interstitial space (4).

Under physiological conditions, most of the oxygen (about 80%) in the mitochondria of cells is reduced by cytochrome oxidase (without the formation of free radicals). The rest of 10–20% enters further oxidation-reduction reactions in the cytoplasm and mitochondria where the oxygen superoxide anion radical (O^{2-}) is formed. Free radicals (SR) are effectively neutralized by a series of enzymatic and non-enzymatic defense mechanisms that include superoxide dismutase, glutathione peroxidase, catalase, glutathione reductase, ascorbic acid, and tocopherol (5, 6).

Free radicals are molecules or fragments of molecules with an unpaired electron in the outer orbit, which makes them highly reactive with a consequent affinity for binding to other molecules. In order to increase the stability of these molecules, there is a high probability of starting a series of reactions when there is peroxidation of lipid membranes and deoxyribonucleic acid (DNA) with consequent cell damage. Free radicals are unstable molecules, or ions of high reactivity, which in the body enter into chemical reactions with parts of the cell (proteins, lipids, carbohydrates, DNA molecules), thereby leading to biochemical, structural, and functional disorders. In a normal molecule, the nucleus is surrounded by a pair of negatively charged electrons. By removing one electron from the pair, through a process called oxidation, the molecule becomes unstable and destructive (a "radical" molecule is created) (7).

Radicals react with neutral biological molecules, damaging them and creating new radicals (chain reaction). A radical can donate an electron to a non-radical and act as a reducing agent, creating new radicals. A radical can take an electron from a non-radical and thus become an oxidizing agent, again creating new radicals. A radical can "tear off" a hydrogen atom from a C–H bond in an organic molecule, creating "carbon-centered" radicals (which with O_2 make peroxyl radicals $\bullet RO_2$). In this way DNA, cell membranes, and proteins are damaged. Radically induced lipid peroxidation reduces membrane fluidity, changes membrane transport, and damages membrane proteins, receptors, enzymes, and ions (8). In such biochemical reactions, free radicals can damage cellular structures (proteins, lipids, and DNA) and cause cell necrosis or apoptosis. This process is called oxidative stress or disturbance of the redox status (the balance between oxidants and antioxidants) (9).

Acute hypoxia is accompanied by significant production of free radicals. During and after hypoxia-ischemia, during the accumulation of arachidonic acid, more free radicals are created. There are several ways of creating free radicals, including by-products in the synthesis of prostaglandins, by the action of the enzyme cyclooxygenase and lipoxygenase (10). If hypoxia persists, the ability of cells to self-protect is reduced, and enzymatic scavengers (e.g., superoxide dismutase and catalase) may be overwhelmed after the ischemic region is reperfused. Restoring blood flow in the ischemic region to "normal" is potentially even more harmful due to the influx of oxygen, which is a source of free radicals. Free radicals are the main cause of endothelial damage, increased permeability, and brain edema after asphyxia/hypoxia (11).

An excess of reactive oxygen species (ROS), including superoxide anion and nitric oxide, which overload antioxidant systems, induces oxidative stress in the body. The main cellular targets of ROS are membrane lipids, proteins, nucleic acids, and carbohydrates. Extensive evidence suggests that ROS play a key role in the initiation and progression of various childhood diseases. Oxidative stress in the newborn has a significant role in the onset and development of retinopathy of prematurity, chronic lung disease, necrotic enterocolitis, renal failure, hemolytic anemia, septic shock, and brain and liver damage. It is alarming to know that just a few minutes of exposure of a newborn to 100% oxygen increases the risk of malignancy in childhood threefold (primarily leukemia) (12).

Hyperoxia (oxidative stress) has an important effect on circulation, especially in the brain and lungs. Although the results are still not clear enough, there is evidence that hyperoxia can cause vasoconstriction in the brain. The same effect can occur in the pulmonary circulation, exacerbating pulmonary hypertension. Examination of a group of newborns with circulatory collapse showed that the recovery of circulation is better if 21% oxygen is applied during resuscitation instead of 100%. Newer findings indicate that tissue damage by supraphysiological amounts of free radicals in newborns occurs during the reoxygenation application of 100% oxygen. Oxidative stress after resuscitation with 100% oxygen lasts for several weeks. In an attempt to discover the initial toxic concentration of oxygen ("threshold"), research has shown that the toxicity of oxygen increases in proportion to the increase in concentration. The risk for the development of oxidative stress depends on the antioxidant capacity. Based on this fact, a full-term newborn has a higher risk than an adult, and a preterm newborn has a significantly higher risk than a full-term newborn (13).

4.1.2 Delayed Effects of Asphyxia

Lack of oxygen causes a series of disorders of organs and organ systems with the development of metabolic acidosis in tissues. The initial metabolic response to lack of oxygen in the tissues is the activation of anaerobic glycolysis in the tissues. This process marks the rapid consumption of the newborn's limited glucose reserves. As a result of the metabolic block, the concentration of lactate increases (14). Ketone bodies are the name given to three compounds (acetone, acetoacetic acid, and beta-hydroxybutyric acid) soluble in water, which are formed as by-products of the use of fatty acids for energy production in the liver and kidney. Ketone bodies are used as a source of energy in the heart and brain and are a vital source of energy for the brain during hypoxia (15).

Disruption of membrane functions, especially changes in the integrity of the plasma membrane, is a very important and probably critical component in acute hypoxia. It is certain that the loss of cell membrane function represents the primary event in the genesis of cell damage in these conditions, but the definitive basis of the mechanisms that cause membrane dysfunction remains unclear. However, it is necessary to emphasize the disturbance of ion homeostasis that occurs in conditions of hypoxia (1).

Hypoxia causes disruption of the calcium gradient and its intracellular accumulation. There is data that in hypoxic conditions the balance of other ions, such as K^+, Na^+, and Cl^-, is disturbed. Hypoxia prevents the flow of potassium through the cell membrane, reducing its net flux, and also changes the function of several types of K^+ channels, especially in nerve cells. Hypoxia causes an increase in intracellular sodium (1, 3).

Cl^- transport takes place via Cl^--stimulated ATPase. In conditions of hypoxia, it is possible to increase intracellular water and the appearance of edema in organs that are dependent on ATP production, where the main cause is the incomplete activity of Na^+ K^+- ATPase. Accumulation of sodium in the cell during ischemia is the immediate cause of intracellular water accumulation (3). In recent research, it is said that extracellular chlorine improves mitochondrial function and has a cardioprotective effect in hypoxia-reperfusion conditions (16).

Hypoxia causes disruption of the calcium gradient and leads to intracellular accumulation. However, today there is a dilemma about how to increase intracellular free calcium in hypoxic conditions. Possible ways are an increase of calcium influx due to the opening of voltage-dependent calcium channels (depolarization) and agonist of operational calcium channels (release of excitatory amino acids), a decrease of calcium efflux due to inhibition of ATP-dependent pumping from the cell, reduced uptake by mitochondria, and endoplasmic reticulum, especially due to changes in mitochondrial membrane potential (17, 18). Accumulation of highly active, free cytosolic calcium may contribute to morphological and physiological transformations leading to cell death after hypoxia (19, 20).

In hypoxia, ATP is broken down, and hypoxanthines accumulate in tissues, which directly leads to cell death (3). Xanthine dehydrogenase and xanthine oxidase catalyze the transformation of hypoxanthine into xanthine and uric acid in the catabolic pathway of purine nucleotides. Under normal physiological conditions, this reaction is catalyzed mainly by xanthine dehydrogenase and does not produce free radicals. However, in the changed conditions, xanthine oxidase is favored, which generates superoxide anion radical and hydrogen peroxide. Hypoxia induces increased transformation of xanthine dehydrogenase into xanthine oxidase. Also, through the increased amount of free cytosolic calcium, xanthine oxidase already present in the cells is activated (21).

A change in the oxidation-reduction state of the cell under conditions of hypoxia, due to acidosis and the possible production of free radicals, can cause the cross-linking of membrane proteins by the formation of disulfide (S–S) bonds. The result is the aggregation of membrane proteins, which changes the activity of ion channels and other cell membrane proteins. By forming mixed disulfides in the reaction with reduced glutathione (GSH), the SH groups of membrane proteins can also be modified in hydroxyl radical-dependent processes. Modified membrane proteins may be an alternative to lipid peroxidation in the mechanism of irreversible cell membrane damage under these conditions (22).

When it comes to the production of lipid peroxides, after acute hypoxia, there is a significant increase in the level of lipid peroxides in the brain with a maximum after 30 minutes, and at the 60th minute, a value almost identical to the normal level is registered. Lipid peroxidation, especially in nerve cells, is catalyzed by iron released from hemoglobin, transferrin, and ferritin, the release of which is induced by tissue acidosis and free oxygen radicals. Hypoxia increases the production of reactive oxygen metabolites, especially when nerve cells are damaged. Due to acidosis, the release of iron ions from ferritin, transferrin, and hemoglobin is favored (23). Under these conditions, hemoglobin can contribute to the formation of free radicals and lipid peroxidation; however, the iron released from ferroproteins is much more active. Namely, free ferrous iron (Fe^{2+}), which is found as a chelator of certain precipitates, can form free radicals (24).

Under physiological conditions, the terminal enzyme of the mitochondrial respiratory chain, cytochrome oxidase, catalyzes the electronic reduction of oxygen to water. The result is an electrochemical gradient across the inner mitochondrial membrane. Under the changed conditions, a partial reduction of O_2 is possible, resulting in the formation of reduced molecules between oxygen and water, representing the following electrons: 55 peroxide anion radical (O_2) (one electron), hydrogen peroxide (H_2O_2) (two electrons), and hydroxyl radical (OH) (three electrons). A possible way of producing free radicals after acute hypoxia is increased concentration of free intracellular calcium, which induces the generation of free oxygen radicals by enzymatic and non-enzymatic means (25).

It has been found that glutamine can lead to rapid neuronal death that occurs within minutes or hours and is explained by intracellular sodium accumulation. Chlorine ions passively follow sodium, and water also diffuses due to the osmotic gradient, leading to neuronal edema and cell membrane lysis (26, 27). The late death of neurons, which takes several hours (sometimes more than 24 hours), is associated with the action of calcium on the cytosol of cells, the activation of enzyme systems, the creation of free radicals, the dysfunction of mitochondria, and the reduction of energy sources. In a similar way, death occurs in many other tissues, although neurons are the most vulnerable to hypoxic-ischemic lesions (28).

Mitochondrial damage is a central event in a cell affected by hypoxia/ischemia (29). The type of mitochondrial damage determines the type of cell death (necrosis or apoptosis) (30). If the damage is intense, it quickly leads to the shutdown of mitochondrial function, loss of ATP production, and explosive rupture of the nuclear and cytoplasmic membrane, which leads to necrosis and uncontrolled cell death (occurs during hypoxia/ischemia). Apoptosis-controlled cell death occurs in the reperfusion-reoxygenation phase. Necroptosis is a programmed necrosis or inflammatory cell disease (31). If the insult is mild, the neurons initially lose but ultimately restore (repair) the function of their mitochondria. However, damaged mitochondria send a signal for the apoptotic program (by activating Cytochrome c or another intramitochondrial protein such as caspase that fragments DNA) (32).

Apoptosis can be much more pronounced in the immature brain. More recent studies show that in the immature brain, in the regions of the cerebral cortex and basal ganglia, there is a greater number of apoptotic than necrotic cells in a period of more than 7 days after hypoxic damage (33). Under conditions of hypoxia (with consequent transient permeability of mitochondria), stimulated by accumulated arachidonic acid, the incomplete reduction of oxygen significantly increases, which leads to the formation of superoxide oxygen anion (O^{2-}) in mitochondria. Other free oxygen radicals are also formed from the superoxide anion. During hypoxia, many other, different pathways occur, including intermediates in the synthesis of prostaglandins, during the activity of the enzyme cyclooxygenase. Thus, mitochondria greatly contribute to oxidative stress, i.e., a state in which the production of free oxygen radicals is greater than their elimination (34).

Incomplete or transient hypoxia/ischemia does not immediately lead to cell death. A neuron can survive the period of ischemia, recover function, or develop permanent neuronal damage hours or days later (35). During hypoxia/ischemia, prolonged membrane depolarization occurs. This depolarization in some neuronal populations leads to the release of excitatory amino acids, among which glutamic and aspartic acids are the most important. Glutamic acid is the most excitatory amino acid in the brain. It is cytotoxic in neuronal and cerebellar cells. Neuronal cells are activated to secrete glutamic acid by the entry of calcium into the cell and their depolarization in hypoxia. A decrease in ATP leads to the maintenance of an increased concentration of glutamic acid, which further leads to prolonged receptor stimulation (2). Activation of these receptors leads to the entry of calcium into neurons with a detrimental effect on cell survival. When energy production is reduced, the ATP-dependent calcium-sodium pump cannot pump all the calcium out of the cell. The result of the mentioned processes is a significant increase in the intracellular concentration of calcium, which activates four types of enzymes. Phospholipases break down the membrane of lipid cells and release arachidonic oxidation in the layers that comprise the prostate and the skin.

Protein kinase increases the synthesis of nitric oxide (NO), while santhine oxidase produces free radicals. Proteases initiate cytoplasmic degradation and endonuclease DNA cleavage (36). This selective vulnerability (especially of the basal ganglia, brainstem, and sensorimotor cortex) is probably more dependent on the intrinsic properties of neurons (i.e., excitatory receptors, reabsorption pumps) than on the circulation and redistribution of cerebral flow (37). In addition, post-hypoxia, ischemia, reperfusion, and reoxygenation can induce the production of large amounts of nitric oxide (NO) in neonatal brain tissue and cerebral circulation (38).

4.1.3 Multiorgan Post-Asphyxia Damage

Unlike an adult organism, a newborn can tolerate hypoxia better because it can use lactates and ketones as an energy substrate several tens of times faster and more efficiently than an adult can (39). The effects of hypoxic and therefore ischemic damage are temporary (reversible) if the duration of hypoxia is limited. However, when hypoxia is prolonged, the affected cells become irreversibly damaged, that is, necrosis occurs despite reperfusion. All metabolic changes that occur as a result of reversible hypoxic cell damage can cause permanent (irreversible) tissue and cell damage and even death (40).

Circulation disorders that lead to brain damage can also lead to damage to other organs. It is a consequence of the redistribution of fetal (neonatal) circulation from less important to vital organs (brain, heart, adrenal gland) (41). Renal disease occurs with oliguria, hematuria, proteinuria, renal insufficiency, and renal vein thrombosis (42).

Postasphyxial lung disease includes meconium aspiration syndrome, persistent fetal circulation, pulmonary hypertension, and respiratory distress syndrome (RDS). Cardiovascular damage occurs with signs of tricuspid insufficiency, myocardial necrosis, hypotension, or shock. Transient bradycardia and arrhythmias are possible. Gastrointestinal disease can present as necrotic enterocolitis and hepatic dysfunction. Hematological disorders appear as thrombocytopenia and disseminated intravascular coagulopathy. Metabolic dysfunction occurs in the form of metabolic acidosis, hypoglycemia, hypocalcemia, and hyponatremia (43). The CNS is the only system whose damage leaves permanent consequences. All other systems affected by asphyxia in the neonatal period generally recover completely (44).

The tissue of the CNS shows the greatest sensitivity to hypoxia. Thus, in the case of a complete interruption of oxygen to the brain, signs of impaired function of the hemispheres appear in 2.5–3 minutes, in the lower structures of the brain in 10–15 minutes, and in the ganglia of the sympathetic nervous system in 60 minutes. At the same time, the parts that are in the excitable phase suffer faster than those that are in the inhibition phase. Changes in the brain under the influence of hypoxia can be registered by electroencephalography (EEG) and by registering delta and beta waves on the EEG (45).

The occurrence of neuropathological and clinical changes characteristic of hypoxic-ischemic encephalopathy (HIE) requires reaching a certain degree of maturation of the central nervous system when there are processes of myelination and the establishment of special synapses. Therefore, HIE primarily occurs in children born at term. In contrast to the term neonate, the prototype of post-asphyxial brain damage in a premature child is represented by intracranial-intraventricular hemorrhage (46).

It is known that asphyxia can arise from a number of causes before or during childbirth. It has been proven that isolated prepartum asphyxia leads to permanent brain damage even without clinical manifestations in the newborn age. It seems that the occurrence of neonatal HIE most often requires the association of prepartum and intrapartum hypoxemic-ischemic stress (47).

The immediate reaction of systemic blood flow to hypoxemia is acceleration of circulation with tachycardia, then vasoconstriction occurs with a decrease in blood flow through peripheral vascular regions and redistribution of circulation. In this phase, the blood flow is directed toward vital organs (heart, brain, adrenal glands) while the kidneys, digestive tract, and muscles are in a state of ischemia. In the case of long-term, severe hypoxemia, peripheral vasoconstriction burdens the heart muscle and reduces blood flow to the right atrium, which leads to a decrease in cardiac output and systemic hypotension with ischemia of all organs (phase of early circulatory decompensation) (48).

Hypoxic cerebral vasodilatation occurs due to the synthesis of lactate and some prostaglandins, as well as an increase in the extracellular concentration of calcium. The so-called cerebrovascular paralysis occurs when there is a disturbance in the ability to preserve the constant flow of blood through the brain, despite variations in systemic pressure (cerebrovascular autoregulation) (49). The initial metabolic response to lack of oxygen in the tissues is the activation of anaerobic glycolysis in the tissues. This process marks the rapid consumption of the newborn child's otherwise limited glucose reserves. As a result of the metabolic block, the concentration of lactate increases. They can be used to some extent for the energy metabolism of the brain. In the case of prolonged and/or more intense hypoxic-ischemic stress, their catabolism occurs with the accumulation of hypoxanthine in the tissues, which directly leads to cell death (50).

The neuropathological substrate of neonatal HIE is the selective necrosis of neurons, the so-called status marmoratus of the basal ganglia, lesions of the parasagittal parts of the brain, and focal ischemic necrosis in the white matter. Periventricular leukomalacia (PVL) can sometimes occur in full-term infants. Selective neuronal necrosis and changes in the basal ganglia are conditioned by the distribution of glutamine substances, while early primary cerebral hypoperfusion plays a role in the development of parasagittal lesions. Focal necrosis and PVL appear to be the result of the combined action of the wound, primary ischemia, and circulatory and metabolic disturbances (51). Newborns who were exposed to more severe prenatal and/or intrapartum hypoxia after birth have a low Apgar score due to varying degrees of neuromuscular depression. If there is no complete recovery despite appropriate resuscitation, HIE develops with an altered state of consciousness, disruption of neuromotor and reflex activity, and signs of vegetative dysfunction. The neurological syndrome in a large number of newborns is complicated by the occurrence of convulsions (52).

Various forms of convulsions are seen in more than half of infants with HIE. Attacks occur in the first day of life, mostly in the second stage. Neurological syndrome is often compromised by disorders caused by hypoxic-ischemic damage to various organs (acute renal failure, meconium aspiration, persistent pulmonary hypertension, and circulatory insufficiency) (53). Along with numerous neurological symptoms, newborns also experience somatic symptoms of varying degrees of severity after intrauterine hypoxia due to the time of onset. Respiratory disorders, cardiovascular disorders, liver, digestive tract, lung, kidney, and other organ function disorders may occur. Therefore, hypoxia of the fetus and newborn is one of the severe complications of pregnancy and childbirth. Its clinical significance is high due to the high incidence, and the consequences of hypoxia are a general medical and social problem. For the above reasons, the coming hypoxia should be diagnosed as soon as possible, prophylaxis should be carried out, and, if necessary, modern treatment should be applied (54, 55).

4.2 Conclusion

Asphyxia affects all organs and leaves transient and permanent sequelae depending on the moment of asphyxia, the gestational maturity of the newborn, as well as the recognition of acute damage and adequate prevention and therapy.

References

1. Vannucci RC, Towfighi J, Vannucci SJ. Secondary energy failure after cerebral hypoxia-ischemia in the immature rat. *J Cereb Blood Flow Metab*. 2004;24:1090–1097.
2. Broughton BR, Reutens DC, Sobey CG. Apoptotic mechanisms after cerebral ischemia. *Stroke*. 2009;40:e331–e339.
3. Marcelino TB, de Lemos Rodrigues PI, Miguel PM, Netto CA, Pereira Silva LO, Matté C. Effect of maternal exercise on biochemical parameters in rats submitted to neonatal hypoxia-ischemia.*Brain Res*. 2015;1622:91–101.
4. Ancarcrona M, Dypbuct JM, Bonfoco F, Zhivotovsky B, Orrenius S, Lipton SA, Nicoter P. Glutamat induced neuronal death: a succession of necrosis or apoptosis dependong on mitochondrial function. *Neuron*. 1995;15:961–973.
5. Buonocore G, Bracci R, Perrone S, Vento M. Oxygen toxicity in newborns. In: Buonocore G, Bracci R, Weindling M, editors. *Neonatology a practical approach to neonatal diseases*, Milano, Italy: Springer-Verlag; 2012, pp. 242–249.
6. Mirończuk-Chodakowska I, Witkowska AM, Zujko ME. Endogenous non-enzymatic antioxidants in the human body. *Adv Med Sci*. 2017;63:68–78.
7. Lushchak VI. Free radicals, reactive oxygen species, oxidative stress and its classification. *Chem.-Biol. Interact*. 2014;224:164–175.

8. Ye ZW, Zhang J, Townsend DM, Tew KD. Oxidative stress, redox regulation and diseases of cellular differentiation. *Biochim. Biophys. Acta.* 2015;1850:1607–1621.
9. Lefaki M, Chondrogianni N. Redox regulation of proteasome function. *Redox Biol.* 2017;13:452–458.
10. Delivoria-Papadopoulos M. Mechanisms of cerebral injury in perinatal asphyxia and stragies for prevention. *J Pediatr.* 1998;132:30–34.
11. Chan PH, Fishman RA. Transient formation of superoxide radicals in polyunsaturated fatty acids-induced brain swelling. *J Neurochem.* 1980;35:1004.
12. Tsukahara H Biomarkers for oxidative stress: clinical application in pediatric medicine. *Curr Med Chem.* 2007;14:339–351.
13. Saugstad OD, Sejersted Y, Solberg R,Wollen EJ, Bjørås. Oxygenation of the newborn: a molecular approach. *Neonatology.* 2012;101:315–325.
14. Taher M, Leen WG, Wevers RA, Willemsen MA. Lactate and its many faces. *Eur J Paediatr Neurol.* 2016;20:3–10.
15. Bon C, Raudrant D, Golfier F, Poloce F, Champion F, Pichot J, Revol A. Feto-maternal metabolism in human normal pregnancies: study of 73 cases. *Ann Biol Clin (Paris).* 2007;65:609–619.
16. Zhang XG, Zhao L, Zhang Y, Li YY, Wang H, Duan GL, Xiao L, Li XR, Chen HP. Extracellular Cl^-- free-induced cardioprotection against hypoxia/reoxygenation is associated with attenuation of mitochondrial permeability transition pore. *Biomed Pharmacother.* 2017;86:637–644.
17. Dong Y, Zhang W, Lai B, Luan WJ, Zhu YH, Zhao BQ, Zheng P. Two free radical pathways mediate chemical hypoxia-induced glutamate release in synaptosomes from the prefrontal cortex. *Biochim Biophys Acta.* 2012;1823:493–504.
18. Vosler PS, Sun D, Wang S, Gao Y, Kintner DB, Signore AP, Cao G, Chen J. Calcium dysregulation induces apoptosis-inducing factor release: cross-talk between PARP-1- and calpain-signaling pathways. *Exp Neurol.* 2009;218:213–220.
19. Liang X, Samways DS, Wolf K, Bowles EA, Richards JP, Bruno J, Dutertre S, DiPaolo RJ, Egan TM. Quantifying Ca^{2+} current and permeability in ATP-gated P2X7 receptors. *J Biol Chem.* 2015;290:7930–7942.
20. Duan Y, Gross RA, Sheu SS. Ca^{2+}-dependent generation of mitochondrial reactive oxygen species serves as a signal for

poly(ADP-ribose) polymerase-1 activation during glutamate excitotoxicity. *J Physiol.* 2007;585:741–758.

21. Szczapa-Krenz H, Gadzinowski J, Szymankiewicz M, Breborowicz GH, Hermann T. Clinical value of hypoxanthine concentration assays in diagnosis of hypoxia in newborns. *Przegl Lek.* 2002;59 Suppl 1:73–77.
22. Farber JL. Mechanisms of cell injury by activated oxygen species. *Environ Health Perspect.* 1994;102 Suppl 10:17–24.
23. Harned J, Nagar S, McGahan MC. Hypoxia controls iron metabolism and glutamate secretion in retinal pigmented epithelial cells. *Biochim Biophys Acta.* 2014;1840:3138–3144.
24. Xu KY, Li F. Iron metabolism and neonatal hypoxic ischemic brain damage. *Sheng Li Xue Bao.* 2017;69:218–224.
25. Mihajlović Lj, Mihajlović N, Petrović A, Pavlović R. Molekularni mehanizam ćelijskih oštećenja pri akutnoj hipoksiji. *Acta medica Medianae.* 2002; 41: 51–61.
26. Rothman SM, Olney JW. Glutamate and the pathophysiology of hypoxic--ischemic brain damage. *Ann Neurol.* 1986;19:105–111.
27. Mayer ML. The challenge of interpreting glutamate-receptor ion-channel structures. *Biophys J.* 2017;113:2143–2151.
28. Rousseaux CG. A review of Glutamate receptors II: pathophysiology and pathology. *J Toxicol Pathol.* 2008;21:133–173.
29. Scorrano L, Oakes SA, Opferman JT, Cheng EH, Sorcinelli MD, Pozzan T, Korsmeyer SJ. BAX and BAK regulation of endoplasmic reticulum Ca^{2+}: a control point for apoptosis. *Science.* 2003;300:135–139.
30. Rousset CI, Baburamani AA, Thornton C, Hagberg H. Mitochondria and perinatal brain injury. *J Matern Fetal Neonatal Med.* 2012;25 Suppl 1:35–38.
31. Graham SH, Chen J. Programmed cell death in cerebral ischemia. *J Cereb Blood Flow Metab.* 2001;21:99–109.
32. Thornton C, Hagberg H. Role of mitochondria in apoptotic and necroptotic cell death in the developing brain. *Clin Chim Acta.* 2015;451(Pt A):35–38.
33. Lockshin RA, Zakeri Z. Caspase-independent cell deaths. *Curr Opin Cell Biol.* 2002;14:727–733.
34. Abe K, Aoki M, Kawagoe J, Yosida T, Hattori A, Kogure K, Itoyan Y. Ischemic delayed neuronal death, a mitochondrial hypothesis. *Stroke.* 1995;26:1478–1489.

35. Torres Cuevas I, Parra-Llorca A, Sánchez-Illana A, Nuñez-Ramiro A, Kuligowski J, Cháfer-Pericás C, Cernada M, Escobar J, Vento M. Oxygen and oxidative stress in the perinatal period. *Redox Biol.* 2017;12:674–681.
36. Johnston MV. Neurbiology of hypoxic-ischemic injury in the developing brain. *Pediatr Res.* 2001;49:735–741.
37. Johnston MV. Selective vulnerability in the neonatal brain. *Ann Neurol.* 1998;44:155–156.
38. Chevalier M, Lory P, Mironneau C, Macrez N, Quignard JF. T-type CaV3.3 calcium channels produce spontaneous low-threshold action potentials and intracellular calcium oscillations. *Eur J Neurosci.* 2006;23:2321–2329.
39. McCord JM. Oxygen-derived free radicals in postischemic tissue injury.*N Engl J Med.* 1985;312:159–163.
40. Martin LJ. The apoptosis-necrosis cell death continuum in CNS development, injury and disease: contributions and mechanisms. In: Lo EH, Marwah J, editors. *Neuroprotection.* Boston: Prominent Press;2002, pp. 379–412.
41. Fellman V, Raivo KO. Reperfusion injury as the mechanism of brain damage after perinatal asphyxia. *Pediatr Res.* 1997;41:599–606.
42. Farber JL, Chien KR, Mittnacht S. The pathogenesis of irreversible cell injury in ischemia. *Am J Pathol.* 1981;271–281.
43. Timovska SN, Cekovska S, Tosheska-Trajkovska K. Acute kidney injury in newborns. *Pril (akedon Akad Nauk Umet Odd Med Nauki).* 2015;36:83–89.
44. Karlo J, Vishnu Bhat B, Koner BC, Adhisivam B. Evaluation of renal function in term babies with perinatal asphyxia. *Indian J Pediatr.* 2014;81:243–247.
45. Hankins GD, Koen S, Gei AF, Lopez SM, Van Hook JW, Anderson GD. Neonatal organ system injury in acute birth asphyxia sufficient to result in neonatal encephalopathy. *Obstet Gynecol.* 2002;99:688–691.
46. Shah P, Riphagen S, Beyene J, Perlman M. Multiorgan dysfunction in infants with post-asphyxial hypoxic-ischemic encephalopathy. *Arch Dis Child Fetal and Neonatal.* 2004;89:152.
47. Jose A, Matthai J, Paul S. Correlation of EEG, CT, and MRI brain with neurological outcome at 12 months in term newborns with hypoxic ischemic encephalopathy. *J Clin Neonatol.* 2013;2:125–130.

48. McAllister JP, Guerra MM, Ruiz LC, Jimenez AJ, Dominguez-Pinos D, Sival D, den Dunnen W, Morales DM, Schmidt RE, Rodriguez EM, Limbrick DD. Ventricular zone disruption in human neonates with intraventricular hemorrhage. *J Neuropathol Exp Neurol.* 2017;76:358–375.
49. Pasternak FJ. Hypoxic-ishemic brain damage in the term infant. *Pediatr. Clin North Am.* 1993;40:1061–1083.
50. Vannucci RC. Hypoxic-ischemic encephalopathy. *Am J Perinatol.* 2000;17:113–120.
51. Hatherill M, Mcintyre AG, Wattie M, Mudoch A. Early hyperlactionaemia in criticaly children. *Intensive Care Med.* 2000;26:314–318.
52. Saugstad SD. Hypoxantine as an indicator of hypoxia. Its role in health and disease through free radical production. *Pediatr Res.* 1988;23:143–150.
53. Hill A. Current concepts of hypoxic-ishaemic cerebral injuri in the term new borns. *Pediatr Neurol.* 1991;5:317–325.
54. Chicwick LM. Hypoxic-ishaemical encephalopathy of the newborn. *Inter Child Heath.* 1994;5:69–75.
55. Sarnat BH, Sarnat SM. Neonatal encephalopathy folowing fetal distress. *Arch. Neurol.* 1976;33:696–701.

Chapter 5

Pathogenesis of Ischemia-Reperfusion Brain Damage

Miljana Z. Jovandaric

5.1 Introduction

Although it makes up only 2% of the total body mass, the brain receives 20% of the cardiac output and requires a constant supply of oxygen and glucose, so the energy needs are seven times higher than in other organs. Pathogenesis of ischemic brain damage involves three phases: ischemic depolarization (ID), biochemical cascade, and reperfusion damage (1). ID leads to interruption of oxygen and glucose supply, resulting in conversion to anaerobic metabolism and accumulation of lactic acid. A few minutes after the arrest, neuronal electrical activity stops and energy loss and ionic homeostasis damage occur. The insufficiency of the membrane ATP-dependent ion pump is followed by the exit of cellular K^+, and the entry of Na^{2+}, Cl^-, and H_2O into the cell (2).

DOI: 10.1201/9781003346692-5

Depolarization of the presynaptic membrane, which happens quickly, damages the transport of neurotransmitters, and a biochemical cascade begins. If the interruption of circulation lasts longer than 5–10 minutes, irreversible cellular changes occur (3, 4). A depolarized cell's presynaptic membrane massively releases the excitatory neurotransmitter glutamate. The accumulation of glutamate in the extracellular space on the postsynaptic membrane leads to the activation of glutamate receptors (NMDA, AMPA, kainate) and of the ion channel complex, which affects the movement of calcium from the extracellular to the intracellular space. The sudden release of glutamate and excessive entry of Ca^{2+} into cells is an excitotoxic process responsible for the cascade of secondary ischemic neuronal damage. A high level of intracellular calcium activates degradative enzymes: proteases, phospholipases, NO synthetase (NOS), calpain (cytosol proteinase), and gelsolin (caspase enzyme), which directly or indirectly damage cellular proteins and other structures leading to accumulation of arachidonic acid, inhibition of mitochondrial respiratory enzymes, and formation of free oxygen radicals and excitotoxic cell death. There is evidence that, in addition to the accumulation of glutamate and intracellular Ca^{2+}, caspase is responsible for apoptotic cell death as the main mediator. NO inhibits apoptosis by inhibiting caspase enzymes (5, 6).

Damage caused by reperfusion and reoxygenation occurs after the establishment of systemic perfusion. Recovery of heart rhythm and stabilization of arterial blood pressure in the period after cardiac arrest does not mean recovery of cerebral perfusion, and cell destruction continues. Changes in cerebral perfusion and pathophysiological consequences of "no-flow" state during and "reflow" state after global cerebral ischemia are complex and insufficiently known. The "no-reflow" phenomenon after cerebral ischemia is the main limiting factor for adequate recovery and survival after cardiac arrest. The stage of damage by reperfusion and reoxygenation is a multifactorial, complex chemical cascade that takes place under the influence of activated leukocytes and cytokines (7).

Leukotrienes and prostaglandins formed from arachidonic acid and free oxygen radicals increase microvascular permeability, which leads to transendothelial migration of leukocytes and their infiltration into ischemic brain tissue and activation of the inflammatory reaction. The result is the further release of a large amount of oxygen free radicals (OFR): superoxide anions, hydrogen peroxide, and hydroxyl radicals. OFRs aggravate tissue damage and the biochemical functioning of cells, causing peroxidation of the lipid component of the cell membrane and further affecting cell lipids, proteins, and nucleic acids, which enables the invasion and infiltration of neutrophils into the post-ischemic brain tissue. NO synthetase increases NO concentration, and an abnormally high amount of superoxide converts NO to peroxynitrite, which damages the capillary endothelium and not only increases edema but also leads to endothelial protrusions that further block the capillaries (3).

Cytokines are proinflammatory and key mediators in the reperfusion phase of cerebral tissue damage. An important source of cytokines is leukocytes. Cytokines induce a further increase in neutrophils that worsen inflammatory activities and lead to necrosis of cells of the nervous system (8). Neuronal damage after hypoxic-ischemic (HI) insult is a biochemical process that includes two phases. Primary energy loss occurs due to the decrease in cerebral flow and oxygen supply. It initiates a cascade of events involving the release of excitatory neurotransmitters and the entry of sodium ions into the cell leading to cytotoxic edema and cell necrosis. After the reperfusion phase, there is a transient recovery of cerebral oxidative metabolism lasting 6 to 12 hours, which represents a window for neuroprotective interventions. Intracellular calcium accumulation leads to further damage. The secondary phase follows, or the phase of delayed neuronal death, which begins 6 to 72 hours after the insult and can last more than 7 days. The production of free radicals and nitric oxide increases with the consequent formation of peroxynitrite, which, along with the activation of inflammatory mediators, initiates mitochondrial dysfunction and cell apoptosis. During the last two decades, studies have shown the critical role of glutamate as a mediator of neuronal death in hypoxic-ischemic lesions (9). According to the location and extent of the damage, ischemia can be divided into two subtypes: focal and global cerebral ischemia (10).

5.1.1 Metabolic Changes in Cerebral Ischemia

Excitotoxicity is the first cellular disorder in cerebral ischemia. The reason is the failure to maintain metabolic homeostasis and excreted metabolites, including glutamate, as the driving force behind ischemia-induced excitotoxicity. It is the most abundant free amino acid in the CNS, which acts as an excitatory neurotransmitter, followed by glycine, D-serine, and polyamines (11, 12). Metabolites associated with excitotoxicity bind to the N-methyl-D-aspartate (NMDA) receptor, which is an ion channel characterized by being permeable to calcium ions and a ligand for driving disturbances in calcium homeostasis in neurons. It can be said that an explicit increase in the excitotoxic neurotransmitter glutamate disrupts the glutamate/glutamine coupling balance, stressing the metabolic homeostasis between astrocytes and neurons (13). The imbalance affects mitochondria and endoplasmic reticulum, which contain Ca^{2+} ion channels. When Ca^{2+} ions massively accumulate in the mitochondria, they disable the mitochondrial function and lead to a decrease in the production of adenosine triphosphate (ATP) and the creation of reactive oxygen species (ROS) (14). During severe mitochondrial dysfunction, the death of neuronal cells occurs through necrosis and oxidation of cellular components such as proteins, lipids, RNA, and

DNA, as well as through apoptosis with the activation of the caspase cascade (15, 16).

Oxidative stress is also caused by excessive production of oxygen derivatives and metabolic dysfunction. Under normal conditions, reactive oxygen species (ROS) are generated in the mitochondrial respiratory chain to produce ATP, and superoxide is converted to hydrogen peroxide, as a signaling messenger, by the enzyme superoxide dismutase. However, in cerebral ischemia, lack of oxygen leads to anaerobic glycolysis in cells, producing lactate as an end product, which in turn makes the cytosolic environment acidic (17). Excess protons convert oxygen into hydrogen peroxide and the reactive hydroxyl radical. As a reflection of these metabolic changes in the brain, increased levels of formate, glycolate, lactate, and pyruvate have been reported in the plasma and urine of patients with cerebral ischemia (18). In addition, oxidative stress associated with metabolic changes has been suggested in the profiles of amino acids, organic acids, free fatty acids, and polyamines in the serum or plasma of patients (19–22).

There are other metabolic changes that occur during cerebral ischemia. Increased levels of formate and glycolate and decreased levels of dimethylamine, glycine, hippurate, and methanol have been suggested to be associated with hyperhomocysteinemia (23, 24). Activation of one-carbon metabolism, including the folic acid cycle and the methionine cycle, is believed to alter the levels of folic acid, cysteine, S-adenosyl homocysteine, and oxidized glutathione, and these changes may represent biomarkers in the serum of patients with cerebral ischemia (25). Altered levels of phosphorus-containing compounds, ceramides, amino acids, fatty acids, carbohydrates, choline, polyamines, and membrane lipids have been proposed as biomarkers for cerebral ischemia, but the lack of a detailed mechanistic explanation for these changes is a limitation (26). Elevated branched-chain amino acids (BCAAs) are known to be catabolized via the ketogenic pathway for energy generation when nutrient availability is scarce in a strong inflammatory cascade and oxidative stress. In patients with cerebral ischemia, BCAA levels have been reported to be decreased due to inflammation. In addition, decreases in lysophosphatidylcholine and polyunsaturated free fatty acids and increases in short-chain free fatty acids, short-chain acylcarnitines, and ceramide were found in the plasma of cerebral ischemia patients with inflammation and β- by oxidation, but by a metabolic mechanism, these findings remain hypothetical (27).

5.1.2 Inflammatory Changes in Cerebral Ischemia

The inflammatory response is also exacerbated by oxidative stress and blood–brain barrier damage in cerebral ischemia. Neuroimmune inflammation is an important component of the pathophysiology of ischemia-reperfusion injuries. Microglia are the first immune cells affected by ischemic brain damage because of their role in the brain by recognizing dead cells and pathogens. Microglia are sensitive to ischemia, and 12 hours after ischemic stroke, 11b-positive (CD11b) microglia in the infarcted area begin to fragment (28). After 24 hours, the number of microglia in the infarct is significantly reduced (29, 30), and microglia in the ischemic penumbra are activated with high expression of CD11b, CD45, and cloned calcium-binding adapter molecule I (lba1). Peri-infarct microglial activation persists for weeks after ischemic stroke (29, 31). In particular, microglia surrounding an ischemic brain lesion display distinct pro- and anti-inflammatory phenotypes (32). However, in experimental models of ischemic stroke, microglia do not appear to exhibit typical M1 and M2 phenotypes. At later stages, microglia aid in stroke recovery by phagocytosing dead cells and debris (33). However, microglia can also phagocytose surviving ischemic neurons, possibly increasing neuronal cell death around the infarct. Infiltrating leukocytes, including neutrophils and monocytes/macrophages, play a complex role in ischemic stroke. Infiltration of neutrophils begins early in the ischemic phase, which binds to endothelial cells through various adhesion molecules (34). On the one hand, neutrophils exacerbate neuronal death by releasing neurotoxic proteases (35). The recruitment of monocytes after the ischemic process is regulated by adhesion molecules, chemokines, and cytokines. Lymphocytes are also closely related to ischemic brain lesions, and there is more and more evidence that subgroups of T cells are involved in the disorder of the regulation of neuroimmune inflammatory functions in the acute phase of the ischemic process (36). After the ischemic process, microglia are rapidly activated, and activated microglia release high levels of IL-23 1 day after ischemic reperfusion, which leads to the infiltration of γδ T lymphocytes into the brain, which worsens brain damage (37).

Autophagy is the mechanism responsible for the continuous removal of unnecessary organelles or misfolded proteins in lysosomes. According to the method of delivery of cellular debris to lysosomes, autophagy is divided into three types: macroautophagy, microautophagy, and chaperone-mediated autophagy. Macroautophagy is the most common form of autophagy, involving the following four steps: (1) initiation of autophagy, (2) vesicle nucleation, (3) vesicle elongation, and (4) fusion and degradation. Through the above processes, autophagy meets the metabolic and restorative demands of cells, recycling denatured organelles and misfolded proteins for reuse to maintain cellular homeostasis (38).

Reperfusion can cause more severe secondary brain tissue injuries than ischemia alone. In 2020, Sun Rui et al. established a mouse model of ischemic stroke via middle cerebral artery occlusion (39). Simultaneously, primary cortical neurons were extracted and cultured, and oxygen and glucose deprivation were used to establish an in vitro model. Their study showed that the low-density lipoprotein receptor inhibits neuronal death after cerebral I/R injury by inhibiting NLRP3-mediated neuronal pyroptosis. Similarly, other studies have also shown that hispidulin, glycosides, remimazolam, and dendrobium alkaloids may have neuroprotective effects in cerebral I/R injury by inhibiting NLRP3-mediated pyroptosis (40, 41). In addition, it has been reported that valproic acid and CHRFAM7A overexpression as well as upregulation of X-box binding protein l (KSBP-1) can also alleviate cerebral I/R injury by inhibiting the canonical caspase-1 pyroptosis pathway (42, 43).

5.2 Conclusion

Knowledge of the process of brain tissue damage during the process of reperfusion after ischemia, with the accumulation of metabolites at the site of the lesion, as well as the inflammatory response of the organism, opens up perspectives for the adequate care of this group of newborns.

References

1. Alvarez-Díaz A, Hilario E, de Cerio FG, Valls-i-Soler A, Alvarez-Díaz FJ. Hypoxic-ischemic injury in the immature brain: key vascular and cellular players. *Neonatology*. 2007;92:227–235.
2. Rainaldi MA, Perlman JM. Pathophysiology of birth asphyxia. *Clin Perinatol*. 2016;43:409–422.
3. Collard CD, Gelman S. Pathophysiology, clinical manifestations and prevention of ischemia-reperfusion injurv. *Anesthesiology*. 2001;94:1133–1138.
4. Hansen KB, Yi F, Perszyk RE, Menniti FS, Traynelis SF. NMDA receptors in the central nervous system. *Methods Mol Biol*. 2017;1677:1–80.

5. Lockshin RA, Zakeri Z. Caspase-independent cell deaths. *Curr Opin Cell Biol.* 2002;14:727–733.
6. Wang X, Karlsson JO, Zhu C, Bahr BA, Hagberg H, Blomgren K. Caspase-3 activation after neonatal rat cerebral hypoxia-ischemia. *Biol Neonate.* 2001;79:172–179.
7. Orrenius S, Zhivotovsky B, Nicotera P. Regulation of cell death: the calcium-apoptosis link. *Nat Rev Mol Cell Biol.* 2003;4:552–565.
8. Guan T, Song J, Wang Y, Guo L, Yuan L, Zhao Y, Gao Y, Lin L, Wang Y, Wei J. Expression and characterization of recombinant bifunctional enzymes with glutathione peroxidase and superoxide dismutase activities. *Free Radic Biol Med.* 2017;110:188–195.
9. Dong Y, Zhang W, Lai B, Luan WJ, Zhu YH, Zhao BQ, Zheng P. Two free radical pathways mediate chemical hypoxia-induced glutamate release in synaptosomes from the prefrontal cortex. *Biochim Biophys Acta.* 2012;1823:493–504.
10. Dawson TM, Dawson VL Mitochondrial mechanisms of neuronal cell death: potential therapeutics. *Annu Rev Pharmacol Toxicol.* 2017;57:437–454.
11. Yao GY, Zhu Q, Xia J, Chen FJ, Huang M, Liu J, Zhou TT, Wei JF, Cui GY, Zheng KY, Hou XY. Ischemic postconditioning confers cerebro protection by stabilizing VDACs after brain ischemia. *Cell Death Dis.* 2018;9:1033.
12. Pinky NF, Wilkie CM, Barnes JR, Parsons MP. Region- and activity-dependent regulation of extracellular glutamate. *J Neurosci.* 2018;38:5351–5366.
13. Huang YH, Sinha SR, Tanaka K, Rothstein JD, Bergles DE. Astrocyte glutamate transporters regulate metabotropic glutamate receptor-mediated excitation of hippocampal interneurons. *J Neurosci.* 2004;24:4551–4559.
14. Doyle S, Hansen DB, Vella J, Bond P, Harper G, Zammit C, Valentino M, Fern R. Vesicular glutamate release from central axons contributes to myelin damage. *Nat Commun.* 2018;9:1032.
15. Doyle KP, Simon RP, Stenzel-Poore MP. Mechanisms of ischemic brain damage. *Neuropharmacology.* 2008;55:310–318.
16. Cai Q, Sun Z, Xu S, Jiao X, Guo S, Li Y, Wu H, Yu X. Disulfiram ameliorates ischemia/reperfusion-induced acute kidney injury by suppressing the caspase-11-GSDMD pathway. *Ren Fail.* 2022;44:1169–1181.

17. Kowalczyk P, Sulejczak D, Kleczkowska P, Bukowska-Ośko I, Kucia M, Popiel M, Wietrak E, Kramkowski K, Wrzosek K, Kaczyńska K. Mitochondrial oxidative stress-a causative factor and therapeutic target in many diseases. *Int J Mol Sci.* 2021;22:13384.
18. Kohen R, Nyska A. Oxidation of biological systems: oxidative stress phenomena, antioxidants, redox reactions, and methods for their quantification. *Toxicol Pathol.* 2002;30:620–650.
19. Holmstrom KM, Finkel T. Cellular mechanisms and physiological consequences of redox-dependent signalling. *Nat Rev Mol Cell Biol.* 2014;15:411–421.
20. Hasanuzzaman M, Alhaithloul HAS, Parvin K, Bhuyan MHMB, Tanveer M, Mohsin SM, Nahar K, Soliman MH, Mahmud JA, Fujita M. Polyamine action under metal/metalloid stress: regulation of biosynthesis, metabolism, and molecular interactions. *Int J Mol Sci.* 2019;20:3215.
21. Chen D, Shao Q, Yin L, Younis A, Zheng B. Polyamine function in plants: metabolism, regulation on development, and roles in abiotic stress responses. *Front Plant Sci.* 2019;9:1945.
22. Schönfeld P, Reiser G. Why does brain metabolism not favor burning of fatty acids to provide energy? Reflections on disadvantages of the use of free fatty acids as fuel for brain. *J Cereb Blood Flow Metab.* 2013;33:1493–1499.
23. Steliga A, Kowiański P, Czuba E, Waśkow M, Moryś J, Lietzau G. Neurovascular unit as a source of ischemic stroke biomarkers-limitations of experimental studies and perspectives for clinical application. *Transl Stroke Res.* 2020;11:553–579.
24. Baranovicova E, Hnilicova P, Kalenska D, Kaplan P, Kovalska M, Tatarkova Z, Tomascova A, Lehotsky J. Metabolic changes induced by cerebral ischemia, the effect of ischemic preconditioning, and hyperhomocysteinemia. *Biomolecules.* 2022;12:554.
25. Kahl A, Stepanova A, Konrad C, Anderson C, Manfredi G, Zhou P, Iadecola C, Galkin A. Critical role of flavin and glutathione in complex I-mediated bioenergetic failure in brain ischemia/reperfusion injury. *Stroke.* 2018;49:1223–1231.
26. Sonderer J, Katan Kahles M. Aetiological blood biomarkers of ischaemic stroke. *Swiss Med Wkly.* 2015;145:w14138.
27. Deleaval P, Luaire B, Laffay P, Jambut-Cadon D, Stauss-Grabo M, Canaud B, Chazot C. Short-term effects of branched-chain amino acids-enriched dialysis fluid on branched-chain amino

acids plasma level and mass balance: a randomized cross-over study. *J Ren Nutr*. 2020;30:61–68.
28. Yang K, Zeng L, Ge A, Wang S, Zeng J, Yuan X, Mei Z, Wang G, Ge J. A systematic review of the research progress of non-coding RNA in neuroinflammation and immune regulation in cerebral infarction/ischemia-reperfusion injury. *Front Immunol*. 2022;13:930171.
29. Lambertsen KL, Meldgaard M, Ladeby R, Finsen B. A quantitative study of microglial-macrophage synthesis of tumor necrosis factor during acute and late focal cerebral ischemia in mice. *J Cereb Blood Flow Metab*. 2005;25:119–135.
30. Ito D, Tanaka K, Suzuki S, Dembo T, Fukuuchi Y. Enhanced expression of Iba1, ionized calcium-binding adapter molecule 1, after transient focal cerebral ischemia in rat brain. *Stroke*. 2001;32:1208–1215.
31. Guruswamy R, ElAli A. Complex roles of microglial cells in ischemic stroke pathobiology: new insights and future directions. *Int J Mol Sci*. 2017;18:496.
32. Clausen BH, Lambertsen KL, Dagnæs-Hansen F, Babcock AA, von Linstow CU, Meldgaard M, et al.. Cell therapy centered on IL-1Ra is neuroprotective in experimental stroke. *Acta Neuropathol*. 2016;131:775–791.
33. Neumann H, Kotter MR, Franklin RJ. Debris clearance by microglia: an essential link between degeneration and regeneration. *Brain*. 2009;132:288–295.
34. Neumann J, Riek-Burchardt M, Herz J, Doeppner TR, König R, Hütten H, et al.. Very-late-antigen-4 (VLA-4)-mediated brain invasion by neutrophils leads to interactions with microglia, increased ischemic injury and impaired behavior in experimental stroke. *Acta Neuropathol*. 2015;129:259–277.
35. Allen C, Thornton P, Denes A, McColl BW, Pierozynski A, Monestier M, et al.. Neutrophil cerebrovascular transmigration triggers rapid neurotoxicity through release of proteases associated with decondensed DNA. *J Immunol*. 2012;189:381–392.
36. Miró-Mur F, Pérez-de-Puig I, Ferrer-Ferrer M, Urra X, Justicia C, Chamorro A, Planas AM. Immature monocytes recruited to the ischemic mouse brain differentiate into macrophages with features of alternative activation. *Brain Behav Immun*. 2016;53:18–33.

37. Komine-Kobayashi M, Chou N, Mochizuki H, Nakao A, Mizuno Y, Urabe T. Dual role of Fcgamma receptor in transient focal cerebral ischemia in mice. *Stroke*. 2004;35:958–963.
38. Tang Y, Xie J, Chen X, Sun L, Xu L, Chen X. A novel link between silent information regulator 1 and autophagy in cerebral ischemia-reperfusion. *Front Neurosci*. 2022;16:1040182.
39. Sun R, Peng M, Xu P, Huang F, Xie Y, Li J, Hong Y, Guo H, Liu Q, Zhu W. Low-density lipoprotein receptor (LDLR) regulates NLRP3-mediated neuronal pyroptosis following cerebral ischemia/reperfusion injury. *J Neuroinflammation*. 2020;17:330.
40. Liu D, Dong Z, Xiang F, Liu H, Wang Y, Wang Q, Rao J. Dendrobium alkaloids promote neural function after cerebral ischemia–reperfusion injury through inhibiting pyroptosis induced neuronal death in both in vivo and in vitro models. *Neurochem Res*. 2020;45:437–454.
41. An P, Xie J, Qiu S, Liu Y, Wang J, Xiu X, Li L, Tang M. Hispidulin exhibits neuroprotective activities against cerebral ischemia reperfusion injury through suppressing NLRP3-mediated pyroptosis. *Life Sci*. 2019;232:116599.
42. Cao X, Wang Y, Gao L. CHRFAM7A overexpression attenuates cerebral ischemia-reperfusion injury via inhibiting microglia pyroptosis mediated by the NLRP3/Caspase-1 pathway. *Inflammation*. 2021;44:1023–1034.
43. Zhang Y, Yao Z, Xiao Y, Zhang X, Liu J. Downregulated XBP-1 rescues cerebral ischemia/reperfusion injury-induced pyroptosis via the NLRP3/Caspase-1/GSDMD axis. *Mediat Inflamm*. 2022;2022:8007078.

Chapter 6

Lipid Peroxidation during Asphyxia

Miljana Z. Jovandaric

6.1 Introduction

There is information that Leonardo da Vinci was the first to notice macroscopic changes in the arteries that correspond to atherosclerosis, with a note that such lesions were later discovered on Egyptian mummies. The term *atherosclerosis* was first used by Marchand around 1860 to emphasize the existence of atheroma and sclerosis in the wall of the aorta and large blood vessels. The presence of cholesterol in atherosclerotic plaque was first pointed out by Vogel in 1847 (1).

The connection between elevated cholesterol levels in the blood and atheroma was observed at the beginning of this century. The credit for this belongs to Aleksandr Ivanovski, who in 1908 published a paper in which he proved that high-energy and high-protein nutrition in rabbits leads to the development of atherosclerosis. Anichkov established that this results in increased infiltration of fat into the arterial wall.

DOI: 10.1201/9781003346692-6

The significance of this animal model was in the morphological and histochemical similarity of these changes with human atherosclerosis. An epoch-making progress in lipidology was marked in 1985 by Michael Brown and Joseph Goldstein for their pioneering work on the role of the LDL receptor, intracellular cholesterol metabolism, and its homeostasis in the body (2).

Undoubtedly, of great importance for a better knowledge of the structure and metabolism of lipoprotein particles in the blood were great advances in the field of knowledge of their protein component – apolipoprotein (3). In recent years, it has been established that hypertriglyceridemia is a risk factor for coronary heart disease and atherosclerosis (4). Since the pioneering works of Ignjatovski and Aničkov at the beginning of this century, great progress has been made and great discoveries have been made that have expanded our knowledge of lipid and lipoprotein metabolism to unimagined limits (5, 6).

6.2 Classification of Lipids

The term *lipids* refers to a group of organic chemical substances that are part of all animal and plant organisms. There are several types of lipids in the human body, among which the most important are fatty acids, cholesterol, triglycerides, and phospholipids. They are an integral part of all complex lipids (cholesterol esters, triglycerides, and phospholipids) and are present in the body in the form of so-called free fatty acids (FFA), unbound to other lipids. Fatty acids are composed of a straight chain of carbon atoms, the number of which is usually even, and the molecule ends with a carboxyl group. According to the number of carbon atoms, they are divided into short- (up to 8 carbon atoms), medium- (8–12 carbon atoms), and long-chain (more than 12 carbon atoms) fatty acids. Based on the presence or absence of double bonds, fatty acids are divided into saturated and unsaturated. Unsaturated is divided according to the number of double bonds into monosaturated (having only one bond) and poly-unsaturated, having two to six double bonds. Unsaturated fatty acids can appear in two isomeric forms, the so-called cis or trans-form (7).

6.2.1 *Fatty Acids*

Fatty acids are found in the blood in two forms, free fatty acids, or non-esterified FFA, and esterified FFA. Esterified fatty acids represent the largest part of the circulation, about 95%, while 5% of free fatty acids are found in the blood. Free fatty acids in plasma are reversibly bound to proteins, primarily to albumins, to a lesser extent to globulins and lipoproteins. They are present in a very low concentration (about 0.5 mmol/l), but they have an extremely fast "turnover" – their half-life is 1–3 minutes. Metabolically, they are the most active fraction of lipids in the plasma, and their oxidation is the main source of energy in the state of starvation. Esterified fatty acids are esterified in the composition of triglycerides (45%), cholesterol esters (15%), and phospholipids (35%) (8).

A large number of fatty acids have been discovered in nature, of which a large amount is present in the human body. Some of the fatty acids can be created in our body, and those fatty acids that the human body is not able to synthesize must be ingested with food and are called essential fatty acids. It has already been mentioned that most of fatty acids are in bound form, in the composition of phospholipids, triglycerides, and cholesterol esters, and only 5% of FFA are in free or unbound form. It should be borne in mind that many properties of triglycerides, cholesterol, and phospholipids depend to a significant extent on the type of fatty acids included in their composition. It is also of great importance whether the fats in the diet are dominated by saturated, monounsaturated, or polyunsaturated fatty acids (9).

6.2.2 Cholesterol

Cholesterol is the best known of the lipid fractions and plays the most important role in the development of atherosclerosis. It is a special type of lipid that differs significantly from other lipid substances in terms of its chemical structure. In its pure form, it is a whitish, soft, waxy substance insoluble in water. It is present in the body in a free, non-esterified, and esterified forms bound with one fatty acid in the form of cholesterol esters. Cholesterol esterification occurs in the plasma under the action of the enzyme lecithin cholesterol-acyltransferase (LCAT). About 75% of cholesterol is found in the plasma in an esterified state, and most often it is esterified with polyunsaturated linoleic fatty acid about 55%. The largest part of free (non-esterified) cholesterol is found in the tissues. Cholesterol is a necessary component of the body. It is a structural element of all cellular and intracellular membranes, and in certain organs it has specific roles (e.g., in hepatocytes it participates in the synthesis of bile acids, in the cortex of the sex and adrenal glands it participates in the synthesis of steroid hormones, and it has the role of a transporter of liposoluble vitamins A, D, E, and K) (10). The origin of cholesterol in the body is twofold (endogenous and exogenous). Most cells have the ability to synthesize it themselves, and its second source is food. It was established that two-thirds of cholesterol is produced by synthesis in the body (about 800–900 mg/day in adults), and only one-third is ingested through food. Considering the ability of the organism to create it in large quantities, it is quite enough to take 150–300 mg per day with food. Excess cholesterol is removed from the body through bile (conversion into bile acids), peeling off the skin, and a very small amount is lost in the urine. Breastfeeding women also lose some cholesterol through their milk. From all of the above, it is obvious that cholesterol is necessary for the body and that it is of great importance for the normal functioning of every cell. Its harmful effects are manifested when it is present in significantly higher concentrations in the blood (11).

6.2.3 *Triglycerides*

Triglycerides are esters of glycerol, a trihydroxyl alcohol with fatty acids. Esterification of only one, two, or all three hydroxyl groups of glycerol with fatty acids is possible, whereby mono-, di-, or triglycerides are formed. Triglycerides are the most widespread in the body, while diglycerides, and especially monoglycerides, are present in significant quantities only in the intestinal mucosa during fat absorption (12), which provides the energy necessary for the life of all cells. The largest amount of triglycerides is found in the composition of fatty tissue (about 95%), while insignificant amounts are present in the blood (13). Even for triglycerides, it has now been reliably established that their increased amount in the blood plays a significant role in the process of atherosclerosis (14).

6.2.4 *Phospholipids*

The chemical structure of phospholipids is very complex. It represents ester-bonded two molecules of fatty acids to one alcohol, which in its molecule contains phosphoric acid as an integral part. Depending on the alcohol they contain, they are divided into two groups, glycerolphospholipids (containing glycerol) and sphingophospholipids (containing sphingosine). With the exception of adipose tissue dominated by triglycerides, phospholipids are the basic lipids of cell membranes and other cellular structural elements. It should be emphasized that the brain and nervous tissue are the richest in phospholipids (15).

6.3 Lipoproteins

An important property of lipids is their insolubility in water. In order to be able to dissolve in the blood and be transported to all cells of the body, all lipid substances are bound to certain proteins, thus creating particles called lipoproteins. The lipid part of lipoprotein particles consists of cholesterol, cholesterol esters, triglycerides, and phospholipids, and their protein parts are very different structures and are called apolipoproteins. Depending on the type and amount of lipids on the one hand and the amount and type of the protein part, or apolipoproteins, on the other hand, different types of lipoproteins are present in the blood (16).

The four basic types of lipoprotein particles present in the blood of all people are chylomicrons, very-low-density lipoproteins (VLDL), low-density lipoproteins (LDL) and high-density lipoproteins (HDL). Chylomicrons, which have the largest lipoprotein particle in the circulation, should be added to these (17).

Certain lipids (cholesterol, triglycerides, phospholipids) are found in the blood in the composition of all lipoprotein particles, in different amounts, and in combination with other components of lipoprotein particles. Chylomicrons and VLDL contain mainly triglycerides, LDL mainly cholesterol, while HDL particles are the richest in protein, and lipids in phospholipids. Cholesterol in a person's blood represents the sum of cholesterol present in all lipoprotein particles, in chylomicrons, VLDL, LDL, and HDL, although its largest amount (about 70%) is found in LDL particles. Depending on the physiological role of individual lipoprotein particles, an increase in cholesterol in one of them will have different significance for the body. Cholesterol found in LDL particles leads to the development of atherosclerosis, while an increase in cholesterol in HDL particles has a protective effect in relation to the development of atherosclerosis. In addition to chylomicrons, VLDL, LDL, and HDL particles, some other types of lipoproteins are normally present in the blood. Among these lipoprotein types, the so-called lipoprotein Lp(a) is of particular importance. It is a special type of lipoprotein that has been proven to be present in the blood of every person in very minimal concentrations (up to 0.25 g/l). Lipoprotein Lp(a), together with the LDL particle, represents the most serious lipid risk factor for the development of atherosclerosis (18).

6.3.1 Apolipoproteins

Lipids are mostly insoluble in water. They can be found in the plasma in the form of stable complexes if they are bound to certain protein parts called apolipoproteins or apoproteins (apo). Previously, it was thought that the protein part of the lipoprotein particle was only a transport vehicle for lipids. However, it was established that they have other, very important roles in the body. Apoproteins ensure the stability of lipid transport in plasma. They are cofactors of certain enzymes that participate in the metabolism of lipoprotein particles. Their most important function consists of binding to specific receptors on cell membranes, which ensures the entry of lipoprotein particles from the plasma into the cells and their further catabolism. The structure and concentration of individual apolipoproteins in a certain lipoprotein particle are under direct genetic control, unlike the content of serum lipids, which are significantly more influenced by diet and metabolic state (19).

6.3.1.1 Apolipoprotein A-I

Apolipoprotein A-I (Apo A-I) is the main apolipoprotein of HDL particles. About 90% of Apo A-I is found in HDL fractions. In the HDL particle itself, it makes up 65% of the entire protein part. It is present in small amounts in chylomicrons and the VLDL fraction. It is synthesized in the liver and the wall of the small intestine. Its most important role is as an activator of the enzyme LCAT, which enables free cholesterol to be esterified on the surface of HDL particles. In this way, the small discoidal HDL3 particles accept excess cholesterol from the cells and accumulate it in the form of esterified cholesterol, where they are transformed into larger, liver-like, less dense HDL2 particles. Considering that it enables the uptake of free cholesterol and later its esterification, it is logical that it has a protective effect on the atherosclerosis process (20).

6.3.1.2 Apolipoprotein A-II

Apolipoprotein A-II (Apo A-II) represents about 20–30% of the total proteins of HDL particles. It is produced in the liver and the wall of the small intestine. It is considered to have an inhibitory effect on LCAT, and a modulatory effect on the activity of lipoprotein lipase (LPL) and hepatic triglyceride lipase (21).

6.3.1.3 Apolipoprotein B

Apolipoprotein B (Apo B) is present in the human population in two forms: apolipoprotein B-48 (Apo B-48) and apolipoprotein B-100 (Apo B-100). Apo B-48 is synthesized in the wall of the small intestine, which enters into the composition of chylomicrons. Far more important is Apo B-100, which is synthesized mainly in the liver and is part of VLDL, intermediate-density lipoproteins (IDL) that circulate for a short time, LDL particles, and lipoprotein Lp(a). It is characteristic of this apoprotein that it also contains certain amounts of carbohydrates (approximately 5% of the total mass), which chemically classifies it as a glycoprotein. Apo B-100 is secreted primarily in the form of VLDL particles from the liver. Further metabolism of VLDL through IDL creates LDL particles that contain only Apo B-100. The most important role of Apo B-100 consists of the specific binding of LDL particles to cellular receptors in the liver and other tissues. Considering that Apo B-100 is the only apolipoprotein in LDL particles, which is known to have the most atherogenic effect, it is understandable that an increase in the concentration of Apo B-100 in plasma increases atherogenic risk. Due to the fact that only one molecule of Apo B-100 is present in an LDL particle, unlike the amount of cholesterol that can vary in them, the plasma concentration of Apo B-100 is a better indicator of the content of atherogenic lipoprotein particles in the blood than the determination of LDL cholesterol (22, 23).

6.3.1.4 Apolipoprotein C

The group of apolipoprotein C (Apo C) is represented by three different proteins that are synthesized in the liver. In circulation, they are bound to chylomicrons, VLDL, and newly created HDL particles. During the intravascular metabolism of lipoprotein particles, there is an intensive exchange of Apo C. Namely, from HDL particles, these apolipoproteins pass to lipoprotein particles rich in triglycerides, that is, to chylomicrons and VLDL. It is evident that Apo C plays a significant role in the catabolism of triglyceride-rich lipoproteins (24).

6.3.1.5 Apolipoprotein E

Apolipoprotein E (Apo E) is a protein molecule composed of 229 amino acids, which contains 10% of the amino acid arginine, which is why it is called an "arginine-rich protein." The main site of its synthesis is the liver, but it is also produced in other organs: the brain, spleen, lungs, adrenal glands, kidneys, ovaries, and muscles. It has a significant role in the metabolism of various lipoprotein particles: chylomicrons, VLDL, IDL, and HDL that contain Apo E. Apolipoprotein E participates in the reversible transport of excess cholesterol from peripheral tissues to the liver by means of HDL particles, which, when loaded with cholesterol, are incorporated into themselves and Apo E. Apo E also appears to be involved in reparative responses to tissue damage. An increase in its concentration is found at the sites of damage and regeneration of peripheral nerves. It is important in nerve regeneration, growth, and/or differentiation of neurons (25).

6.3.1.6 Apolipoprotein (A)

Apolipoprotein (Apo A) is a relatively high-molecular-weight protein that is part of the lipoprotein Lp (a). If Apo A is separated, the rest of Lp(a) is almost identical in composition to the LDL particle because it has a similar lipid composition and contains one molecule of Apo B-100. The gene that regulates the synthesis of this apolipoprotein is located on the sixth chromosome near the gene for plasminogen. Apo A has a great structural similarity and shows immunological cross-reactivity with plasminogen, which could represent a connection between the process of atherosclerosis and the process of fibrinolysis (26).

6.4 Lipids Transport and Metabolism in the Organism

6.4.1 Transport of Lipids in the Body

During digestion, fats from food are absorbed by lymph in the intestines, with the exception of short-chain fatty acids. Most triglycerides are broken down into monoglycerides and fatty acids in the digestive tract. When passing through the intestinal epithelial cells, monoglycerides and fatty acids are resynthesized into new triglyceride molecules that enter the lymph as small, dispersed droplets called chylomicrons, with a diameter between 0.08 and 0.6 microns. A small amount of protein, apoprotein B, is absorbed on the outer surface of chylomicrons and thus increases the stability of the chylomicron suspension in the lymph fluid, preventing their adhesion to the walls of lymphatic vessels. Most of the cholesterol and phospholipids that are absorbed from the gastrointestinal tract enter the composition of chylomicrons. The basic structure of chylomicrons consists of triglycerides, the composition of which includes about 9% phospholipids, 3% cholesterol, and 1% apoprotein B. Chylomicrons are transported through the ductus thoracicus, which empties into the venous blood at the composition of the jugular vein and subclavian. One hour after a large fatty meal, the concentration of chylomicrons in the plasma can even rise to 1–2%, so due to their large amount, the plasma becomes cloudy and sometimes yellow. However, chylomicrons have a half-life of less than 1 hour, so the plasma becomes clear again within a few hours (27).

Most of the chylomicrons are removed from the blood when they pass through the capillaries of adipose tissue and the liver. Both tissues contain large amounts of an enzyme called lipoprotein lipase. This enzyme acts in the capillary endothelium where it hydrolyzes chylomicron triglycerides. When they come into contact with the wall of the endothelium, it breaks them down into fatty acids and glycerol. Fatty acids, which easily pass through the cell membrane, diffuse immediately into fat cells or liver cells. Once found in these cells, fatty acids are resynthesized into triglycerides, using the glycerol produced in the metabolic processes in the cells. When the fat that is deposited in adipose tissue becomes needed by some other part of the body, usually as a source of energy, it must first be transported from adipose tissue to other tissues. This transfer is mainly carried out in the form of FFA. They are formed by rehydrolysis of triglycerides into fatty acids and glycerol. At least two types of stimuli play an important role in promoting hydrolysis. First, when glucose is very low in fat cells, the level of one of its breakdown products, alpha-glycerophosphate, drops sharply. Since this substance is necessary for the creation of the glycerol part in the process of triglyceride synthesis, its deficiency shifts the balance in the direction of hydrolysis. Second, hormone-sensitive cellular lipase, which can be activated by various hormones, accelerates the hydrolysis of triglycerides (28, 29).

At the moment of leaving fat cells, fatty acids are ionized in the plasma and immediately bind to the albumin fraction of plasma proteins. A fatty acid bound in this way to albumin is called a FFA or non-esterified fatty acid, to distinguish it from other plasma fatty acids that exist as esters of glycerol, cholesterol, or other compounds. Most often, 3 fatty acid molecules are attached to 1 albumin molecule, but when the need to transport fatty acids becomes extreme, up to 30 fatty acid molecules can be attached to 1 albumin molecule (30).

6.4.2 Lipid Metabolism in the Body

The transport and metabolism of lipids in the body consists of three pathways: transport of lipids absorbed in the intestines, transfer of lipids synthesized in the body to the liver, and return of excess cholesterol from peripheral tissues to the liver. Through the mucosa of the small intestine and in the intestinal wall itself, they incorporate chylomicrons into the largest lipoprotein particles. Through the lymphatic vessels, they reach the capillaries, where the chylomicron particles are broken down under the action of lipoprotein lipase. Triglycerides are released from chylomicrons, which enter fat tissue, where they are deposited and serve as an energy reservoir, and into muscles, where they are oxidized and represent a source of necessary energy for muscle work. The remaining parts of chylomicrons, which contain significant amounts of cholesterol (the so-called chylomicron remnants), go to the liver where they are further broken down (31).

The transport route and metabolism of lipids created in the body begins in the liver. Cholesterol, triglycerides, and phospholipids reach the liver or are formed in it, building together with specific proteins large lipoprotein particles, VLDL particles, or very-low-density lipoproteins. They are smaller than chylomicrons and contain more cholesterol. VLDL particles leave the liver and enter the bloodstream, where they release triglycerides in the capillaries that go to muscles and fat tissue. By releasing triglycerides, the size of these particles decreases and the percentage of cholesterol in them increases. LDL particles or low-density lipoproteins are formed from VLDL particles, which contain larger amounts of cholesterol and less triglycerides. It should be mentioned that during the transition to LDL particles, there is a transitional stage, the so-called intermediate lipoprotein particles (IDL), whose life in the circulation is extremely short. The basic physiological role of LDL particles is to ensure the delivery of necessary amounts of cholesterol to every cell of our body (32). In order for LDL particles to penetrate the cell, there are special structural elements called receptors on the cell membrane. It should be emphasized that all the cells of our body can create cholesterol themselves (33).

Cholesterol that has reached the cell has the role of its further accumulation in the cell. When there is a lot of cholesterol in the cell, it defends itself from further accumulation using two mechanisms: it reduces the number of receptors on its surface, which in turn reduces the entry of LDL particles into the cell. Accumulated cholesterol in the cell inhibits the activity of the enzyme hydroxymethylglutaryl-coenzyme

A-reductase, which is essential for the formation of cholesterol in the cell (34).

Receptors for LDL particles are found on all cells. It has been proven that cells that are actively dividing and therefore need more cholesterol to build cell membranes have more of these receptors. Cells of the liver, adrenal glands, and gonads, i.e., organs with extremely high requirements for cholesterol, have the largest number of receptors on their surface (35).

The way in which the cell defends itself against the increased influx of cholesterol is actually an autoregulatory mechanism that prevents excessive accumulation of cholesterol. However, not all cells have such a cholesterol influx control mechanism. There are cells on the surface of which there is another type of receptor, called scavenger receptors, which accept practically unlimited amounts of cholesterol. It is characteristic that such cells include smooth muscle cells in the blood vessel wall, cells lining the inside of the blood vessel (i.e., endothelial cells), and a special type of cell, the so-called macrophages, and all these cells play a significant role in the process of atherosclerosis (36).

It has been proven that under normal circumstances, about 80% of LDL particles are removed from the circulation through specific receptors with a mechanism of autoregulation of cholesterol influx into the cell, and only 20% through the "cleaner" receptor system (37, 38).

The third way of transport and metabolism of lipids is represented by HDL particles or high-density lipoproteins. They are mostly created in the liver and small intestine, but they can also be created in the bloodstream itself. Their basic function consists of accepting excess cholesterol from the cells and transporting it back to the liver. Given that HDL particles enable the removal of excess cholesterol from cells, it is understandable that they have a protective effect in relation to the atherosclerosis process. It has been proven that HDL subfraction has a protective effect (39). The fact that

HDL protects LDL particles from free radicals and their subsequent oxidation is of great importance for the protective effect of HDL particles (40).

6.5 Influence of Hypoxia/Ischemia on Lipids

6.5.1 Free Radicals

Radicals are normal components of cells, and they are formed by the action of oxygen molecules on cell membrane proteins. Under physiological conditions, most of the oxygen (about 80%) in the mitochondria of cells is reduced by cytochrome oxidase (without the formation of free radicals). The rest of 10–20% enters further oxidation-reduction reactions in the cytoplasm and mitochondria where the oxygen superoxide anion radical (O^{2-}) is formed. Free radicals (SR) are effectively neutralized by a series of enzymatic and non-enzymatic defense mechanisms that include superoxide dismutase, glutathione peroxidase, catalase, glutathione reductase, ascorbic acid, tocopherol, and they act as free radical scavengers (41, 42).

Free radicals are molecules or fragments of molecules with an unpaired electron in the outer orbit, which makes them highly reactive with a consequent affinity for binding to other molecules. In order to increase the stability of these molecules, there is a high probability of starting a series of reactions when there is peroxidation of lipid membranes and deoxyribonucleic acid (DNA) with consequent cell damage. Free radicals are unstable molecules, or ions of high reactivity, which in the body enter into chemical reactions with parts of the cell (proteins, lipids, carbohydrates, DNA molecules), thereby leading to biochemical, structural, and functional disorders. In a normal molecule, the nucleus is surrounded by a pair of negatively charged electrons. By removing one electron from the pair, through a process called oxidation, the molecule becomes unstable and destructive (a "radical" molecule is formed) (43).

Radicals react with neutral biological molecules, damaging them and creating new radicals (chain reaction). A radical can donate an electron to a non-radical and act as a reducing agent, creating new radicals. A radical can take an electron from a non-radical and thus become an oxidizing agent, again creating new radicals. A radical can "tear off" a hydrogen atom from a C-H bond in an organic molecule, creating "carbon-centered" radicals (which with O_2 make peroxyl radicals $\bullet RO_2$). In this way, DNA, cell membranes, and proteins are damaged. Radically induced lipid peroxidation reduces membrane fluidity, changes membrane transport, and damages membrane proteins, receptors, enzymes, and ions (44). In such biochemical reactions, free radicals can damage cellular structures (proteins, lipids, and DNA) and cause cell necrosis or apoptosis. This process is called oxidative stress or disturbance of the redox status (the balance between oxidants and antioxidants) (45).

Acute hypoxia is accompanied by significant production of free radicals. During and after hypoxia-ischemia, during the accumulation of arachidonic acid, more free radicals are created. There are several ways of creating free radicals, including by-products in the synthesis of prostaglandins, by the action of the enzyme cyclooxygenase and lipoxygenase (46). If hypoxia persists, the ability of cells to self-protect is reduced, and enzymatic scavengers (e.g., superoxide dismutase and catalase) may be overwhelmed (overwhelmed) after the ischemic region is reperfused. Restoring blood flow in the ischemic region to "normal" is potentially even more harmful due to the influx of oxygen, which is a source of free radicals. Free radicals are a major cause of endothelial damage, increased permeability, and brain edema after asphyxia/hypoxia (47).

An excess of reactive oxygen species (ROS) including superoxide anion and nitric oxide, which overload antioxidant systems, induce oxidative stress in the body. The main cellular targets of ROS are membrane lipids, proteins, nucleic acids, and carbohydrates. Extensive evidence suggests that ROS plays a key role in the initiation and progression of various childhood diseases. Oxidative stress in the newborn has a significant role in the onset and development of retinopathy of prematurity, chronic lung disease, necrotic enterocolitis, renal failure, hemolytic anemia, septic shock, and brain and liver damage. It is alarming to know that just a few minutes of exposure of a newborn to 100% oxygen increases the risk of malignancy in childhood threefold (primarily leukemia) (48).

Hyperoxia (oxidative stress) has an important effect on circulation, especially in the brain and lungs. Although the results are still not clear enough, there is evidence that hyperoxia can cause vasoconstriction in the brain. The same effect can occur in the pulmonary circulation, exacerbating pulmonary hypertension. An examination of a group of newborns with circulatory collapse showed that the recovery of circulation is better if 21% oxygen is applied during resuscitation instead of 100%. Newer findings indicate that tissue damage by supraphysiological amounts of free radicals in newborns occurs during the reoxygenation application of 100% oxygen. Oxidative stress after resuscitation with 100% oxygen lasts for several weeks. In an attempt to discover the initial toxic concentration of oxygen ("threshold"), research has shown that the toxicity of oxygen increases in proportion to the increase in concentration. The risk for the development of oxidative stress depends on the antioxidant capacity. Based on this fact, a term newborn has a higher risk than an adult, and a preterm newborn has a significantly higher risk than a term newborn (49).

Lack of oxygen causes a series of disorders of organs and organ systems with the development of metabolic acidosis in tissues. The initial metabolic response to lack of oxygen in the tissues is the activation of anaerobic glycolysis in the tissues. This process marks the rapid consumption of the newborn child's otherwise limited glucose reserves. As a result of the metabolic block, the concentration of lactate increases (50). Ketone bodies are the name given to three compounds (acetone, acetoacetic acid, and beta-hydroxybutyric acid) soluble in water, which are formed as by-products of the use of fatty acids for energy production in the liver and kidney. Ketone bodies are used as an energy source in the heart and brain and are a vital source of energy for the brain during hypoxia (51, 52).

In hypoxia, adenosine triphosphate (ATP) is broken down, and hypoxanthines accumulate in tissues, which directly leads to cell death (53). Xanthine dehydrogenase and xanthine oxidase catalyze the transformation of hypoxanthine into xanthine and uric acid in the catabolic pathway of purine nucleotides. Under normal physiological conditions, this reaction is catalyzed mainly by xanthine dehydrogenase and does not produce free radicals (54). However, in the changed conditions, xanthine oxidase is favored, which generates superoxide anion radical and hydrogen peroxide. Hypoxia induces increased transformation of xanthine dehydrogenase into xanthine oxidase (55). Also, through the increased amount of free cytosolic calcium, xanthine oxidase already present in the cells is activated (56).

6.5.2 Lipid Peroxidation

A change in the oxidation-reduction state of the cell under conditions of hypoxia, due to acidosis and the possible production of free radicals, can cause the cross-linking of membrane proteins by the formation of disulfide (S–S) bonds. The result is the aggregation of membrane proteins, which changes the activity of ion channels and other cell membrane proteins. By forming mixed disulfides in the reaction with reduced glutathione (GSH), SH groups of membrane proteins can also be modified in hydroxyl radical-dependent processes. Modified membrane proteins may be an alternative to lipid peroxidation in the mechanism of irreversible cell membrane damage under these conditions (57).

When it comes to the production of lipid peroxides, after acute hypoxia, there is a significant increase in the level of lipid peroxides in the brain with a maximum after 30 minutes, and at the 60th minute, a value almost identical to the normal level was registered. Lipid peroxidation, especially in nerve cells, is catalyzed by iron released from hemoglobin, transferrin, and ferritin, the release of which is induced by tissue acidosis and free oxygen radicals. Hypoxia increases the production of reactive oxygen metabolites, especially when nerve cells are damaged. Due to acidosis, the release of iron ions from ferritin, transferrin, and hemoglobin is favored (58). Under these conditions, hemoglobin can contribute to the formation of free radicals and lipid peroxidation; however, the iron released from ferroproteins is much more active. Namely, free ferrous iron (Fe^{2+}), which is found as a chelator of certain precipitates, can form free radicals (59). Under physiological conditions, the terminal enzyme of the mitochondrial respiratory chain, cytochrome oxidase, catalyzes the electronic reduction of oxygen to water. The result is an electrochemical gradient across the inner mitochondrial membrane. Under the changed conditions, a partial reduction of O_2 is possible, resulting in the formation of reduced molecules between oxygen and water, representing the following electrons: 55 peroxide anion radical (O_2) (one electron), hydrogen peroxide (H_2O_2) (two electrons), and hydroxyl radical (OH) (three electrons). A possible way of producing free radicals after acute hypoxia is an increased concentration of free intracellular calcium, which induces the generation of free oxygen radicals by enzymatic and non-enzymatic means (60).

It has been found that glutamine can lead to rapid neuronal death that occurs within minutes or hours and is explained by intracellular sodium accumulation. Chlorine ions passively follow sodium, and water also diffuses due to the osmotic gradient, leading to neuronal edema and cell membrane lysis (61–63). The late death of neurons, which takes several hours (sometimes more than 24 hours), is associated with the action of calcium on the cytosol of cells, the activation of enzyme systems, the creation of free radicals, the dysfunction of mitochondria, and the reduction of energy sources. Death occurs in a similar way in many other tissues, although neurons are the most vulnerable to hypoxic-ischemic lesions (64).

Mitochondrial damage is a central event in a cell affected by hypoxia/ischemia (65). The type of mitochondrial damage determines the type of cell death (necrosis or apoptosis) (66). If the damage is intense, it quickly leads to the shutdown of mitochondrial function, loss of ATP production, and explosive rupture of the nuclear and cytoplasmic membrane, which leads to necrosis and uncontrolled cell death (occurs during hypoxia/ischemia). Apoptosis-controlled cell death occurs in the reperfusion-reoxygenation phase. Necroptosis is a programmed necrosis or inflammatory cell disease (67). If the insult is mild, the neurons initially lose but ultimately restore (repair) the function of their mitochondria. However, damaged mitochondria send a signal for the apoptotic program (by activating Cytochrome c or another intramitochondrial protein such as caspase that fragments DNA) (68).

Apoptosis can be much more pronounced in the immature brain. More recent studies show that in the immature brain, in the regions of the cerebral cortex and basal ganglia, there is a greater number of apoptotic than necrotic cells for a period of more than 7 days after hypoxic damage (69). Under conditions of hypoxia (with consequent transient permeability of mitochondria), stimulated by accumulated arachidonic acid, the incomplete reduction of oxygen significantly increases, which leads to the formation of superoxide oxygen anion (O^{2-}) in mitochondria. Other free oxygen radicals are also formed from the superoxide anion. During hypoxia, many other, different pathways occur, including intermediates in the synthesis of prostaglandins, during the activity of the enzyme cyclooxygenase. Thus, mitochondria greatly contribute to oxidative stress, i.e., a state in which the production of free oxygen radicals is greater than their elimination (70).

Incomplete or transient hypoxia/ischemia does not immediately lead to cell death. The neuron may survive the period of ischemia, and recover function, or permanent neuronal damage may occur hours or days later (71). During hypoxia/ischemia, prolonged membrane depolarization occurs. This depolarization in some neuronal populations leads to the release of excitatory amino acids, among which glutamic and aspartic acids are the most important. Glutamic acid is the most excitatory amino acid in the brain. It is cytotoxic in neuronal and cerebellar cells. Neuronal cells are activated to secrete glutamic acid by the entry of calcium into the cell and their depolarization in hypoxia. The drop in ATP leads to the maintenance of an increased concentration of glutamic acid, which further leads to prolonged receptor stimulation. Activation of these receptors leads to the entry of calcium into neurons with a detrimental effect on cell survival. When energy production is reduced, the ATP-dependent calcium-sodium pump cannot pump all the calcium out of the cell. The result is a significant increase in the intracellular concentration of calcium, which activates at least four types of enzymes: phospholipases (decompose the lipid cell membrane, release arachidonic acid, which creates prostaglandins and free radicals); protein kinase that activates enzymes, including nitric oxide (NO) synthesis, xanthine oxidase that also produces free radicals. Proteases start the degradation of the cytoplasm, and endonucleases the

DNA cleavage (72). This selective vulnerability (especially of the basal ganglia, brainstem, and sensorimotor cortex) is probably more dependent on the intrinsic properties of neurons (i.e., excitatory receptors, reabsorption pumps) than on the circulation and redistribution of cerebral flow (73). In addition, post-hypoxia, ischemia, reperfusion, and reoxygenation can induce the production of large amounts of nitric oxide (NO) in neonatal brain tissue and cerebral circulation (74).

Unlike an adult organism, a newborn can tolerate hypoxia better because it can use lactates and ketones as an energy substrate several tens of times faster and more efficiently (75). The effects of hypoxic and therefore ischemic damage are temporary (reversible) if the duration of hypoxia is limited. However, when hypoxia is prolonged, the affected cells become irreversibly damaged, that is, necrosis occurs despite reperfusion. All metabolic changes that occur as a result of reversible hypoxic cell damage can cause permanent (irreversible) tissue and cell damage and even death (76).

Loss of cell membrane function is the primary event in the occurrence of cell damage under hypoxic conditions, but the definitive basis of the mechanisms that cause membrane dysfunction remains unclear. It is necessary to emphasize the existence of the following potential mechanisms: peroxidation of membrane lipids, cross-linking of membrane proteins, and increased degradation of phospholipids. Hypoxia leads to the oxidation of lipoproteins, where there is an interdependence of the concentration of lipid hydroperoxides in the plasma with a more severe degree of hypoxia, that is, with the manifestation of the consequences of hypoxia. The most commonly oxidized lipids are in low-density lipoproteins (LDL), and LDL is the main source of lipids that accumulate during ischemia (77). Lipid peroxidation is oxidative damage that affects cell membranes, i.e., lipids of membrane lipoproteins, as well as plasma lipids (78).

6.5.3 Peroxidation of Membrane Lipids

In recent years, lipid peroxidation has been the subject of extensive studies in terms of mechanism dynamics, analysis of end products, and association with diseases. Lipids are oxidized by three different mechanisms: enzymatic oxidation, non-enzymatic oxidation, and oxidation mediated by free radicals. The relative sensitivity of lipids to oxidation depends on their structure. Lipid hydroperoxides are formed as primary products, but they are substrates for different enzymes and thus undergo different secondary processes. Different types of antioxidants with different functions inhibit the harmful effects of lipid peroxidation products. The biological role of lipid peroxidation has recently attracted much attention, but its physiological significance must be demonstrated in subsequent studies (79). Lipid peroxidation, especially in nerve cells, is catalyzed by iron released from hemoglobin, transferrin, and ferritin, the release of which is induced by tissue acidosis and free oxygen radicals (80).

In circulation, lipids are packed in lipoprotein molecules because they are insoluble in water. Lipoproteins have a central lipid part (core) in which there are triglycerides and cholesterol esters, and on the surface there is an envelope composed of certain proteins (apoproteins), phospholipids, and a small amount of free cholesterol. Thanks to this envelope, lipids can be transported through the blood. Lipoproteins are large molecular complexes of lipids and apoproteins (apo), which transport hydrophobic plasmatic lipids, mainly cholesterol and triglycerides, in the plasma (81).

Oxidatively modified LDL particles can also be created by direct oxidation, which mainly takes place on unsaturated fatty acids under the influence of superoxide and peroxynitrite radicals, or by the transfer of already oxidized cellular lipids to LDL. In doing so, lipid peroxides are formed, which enter into a new chain reaction, as well as conjugated dienes and aldehyde products. These changes are preceded by a "hesitation period" of aldehyde oxidation, i.e., a latent period in which the antioxidants in LDL are consumed, after which the oxidation of fatty acids can begin (82, 83).

At the same time, degraded LDL becomes a ligand for macrophage receptors, and in this way monocytes are attracted to the site of the lesion, "take" lipoproteins, and gradually turn into foam cells. When LDL deposits exceed a critical threshold, excessive activation of complement and macrophages creates the conditions for chronic inflammation (84).

Activated macrophages cause the death of endothelial cells by the process of apoptosis or the secretion of proteolytic enzymes, and secrete numerous growth factors. The apoptotic effect of Ox-LDL on endothelial cells can be attributed to oxidation products of phosphatidylcholine or oxysterol (85, 86).

6.5.3.1 The Pathophysiological Importance of the Oxidized LDL Particle

Lipid peroxidation (LPO), the most pronounced negative phenomenon in the action of free radicals, represents an autocatalytic, progressive, and most often irreversible process. The initiation of lipid peroxidation process is initiated by any oxidant capable of borrowing a hydrogen atom from the double bond of an unsaturated fatty acid. The most responsible system for the oxidative modification of the LDL particle is the cellular lipoxygenase system (87).

Oxidative modification of LDL is possible through peroxynitrite, which is formed in the reaction of nitrogen monoxide (NO) and superoxide, which are produced to an increased extent in conditions of endothelial activation, considering that there is induction of NO synthase and endothelial NADPH oxidase (88).

Oxidized LDL (Oh-LDL) is a modified form of LDL, where the oxidative modification takes place via SR. Namely, in the course of normal metabolic activity, there is a controlled creation of free radicals (atoms or molecules that have one unpaired electron, which is why they are very unstable and reactive). When metabolic disorders occur, free radicals are created uncontrollably and lead to significant damage to macromolecules. Free radicals react with lipids, proteins, and nucleic acids, leading to their oxidative modification. When lipids are damaged by free radicals, lipid free radicals and lipid peroxides are formed. It is considered that the main agent of LDL oxidation is the superoxide anion, under the influence of which both the lipid and protein components of the LDL particle are modified (89).

LDL differs from unmodified LDL in composition, structure, and metabolic pathway (primarily catabolism). LDL contains a higher percentage of fatty acids and lysophosphatidylcholine as a result of the action of phospholipase A2. Oh-LDL is characterized by reactive aldehydes, reduced cholesterol content with the reduced number of reactive amino groups on apo B-100, elevated oxysterols, and fragmentation of apo B-100. Peroxidation of polyunsaturated fatty acids leads to the formation of highly reactive aldehydes that have a cytotoxic effect. Oxidation of apo B-100 leads to the loss of antigenic tolerance with the appearance of antibodies to modified LDL in the intima of arteries. As a result, there is a reduced recognition of the LDL receptor by the LDL particle and its increased binding to free radicals. There are no autoregulatory mechanisms for this pathway, so the cells that use it are not protected from uncontrolled cholesterol accumulation. Such cells include smooth muscle cells and monocytic macrophage cells (90). Other factors that may contribute to the uptake of LDL by macrophages by receptor or non-receptor mechanisms include phagocytosis, self-aggregation of LDL, formation of complexes with proteoglycans, formation of immune complexes, and degradation by hydrolytic enzymes. Oxidative modification of "trapped" LDL particles takes place in two phases. The first phase occurs before the mobilization of monocytes, and the second after infiltration by monocytes/macrophages, thanks to their high oxidative capacity. Native LDL particles in the subendothelial space can also be oxidized by resident vascular cells: smooth muscle, endothelial, and monocyte/macrophage cells (91).

There are two stages in the oxidative modification of the LDL particle. Minimally oxidized LDL occurs when only the lipid component is slightly oxidized, in contrast to complete oxidation of LDL when there are changes in the structure of apo B-100, which fragments and creates larger molecular aggregates and becomes more electronegative. Modification of 5–10% of lysine residues on apo B-100 changes its structure and leads to the removal of LDL particles via receptors on macrophages. LDL particles have chemotaxis properties for monocytes, inhibition of macrophage migration from the blood vessel wall into the circulation, uptake through specific receptors, and cytotoxic properties (92).

Ox-LDL stimulates endothelial cells to produce monocyte chemotaxis factor, and the result of these events is the accumulation of monocytes in the intima of blood vessels and their differentiation into macrophages. Ox-LDL, due to the presence of modified apo B-100, is not recognized by normal receptors but is recognized with high affinity by irregular SR on activated macrophages. Uncontrolled uptake of Ox-LDL by SR on macrophages leads to their destruction. Macrophages can secrete Ox-LDL and superoxide anion, thereby becoming cytotoxic to the environment and further damaging the endothelium (93).

Activated macrophages cause the death of endothelial cells by the process of apoptosis or the secretion of proteolytic enzymes. The apoptotic effect of Ox-LDL on endothelial cells may be attributed to oxidation products of phosphatidylcholine or oxysterol (94).

Under conditions of hypoxia, oxidation of lipids occurs, especially polyunsaturated fatty acids (PUFA) and cholesterol. Oxysterols as cholesterol oxidation products are a diagnostic biomarker of oxidative stress (95).

The presence of cholesterol in the brain is essential for the maturation of neurons. The most frequently oxidized lipids are LDL. This susceptibility to oxidation can be attributed to the physical properties of the membrane and the outer surface of the nucleus, where the two lipid envelopes are located. The cell membrane is most susceptible to peroxidation due to the high content of unsaturated fatty acids (96).

Lipid peroxides are released when free radicals react with unsaturated fatty acids of the cell membrane during hypoxic-ischemic disease in infants (97). Under hypoxic conditions, after cell damage or loss of cell membrane integrity, lipid peroxidation continues, especially if oxygen therapy is administered. Chemically modified LDL is recognized by a specific macrophage receptor, which leads to the entry of LDL and cholesterol into the cell. In the circulation, lipids undergo oxidation, peroxidation, and removal from the circulation, while cellular lipids are still found locally without entering the circulation (98).

6.6 Lipid Peroxidation in Pregnancy

Pregnancy is a very dynamic process accompanied by numerous metabolic changes in order to adapt to the needs of placenta formation, rapid fetal development, childbirth, and lactation (99). Metabolic changes vary among pregnant women depending on lifestyle habits, health status before and during pregnancy, and fetal growth dynamics, which is reflected in variations in nutrients such as glucose, free fatty acids, unsaturated fatty acids, amino acids, vitamins, and minerals needed for normal fetal development (100).

The growth of the fetus is influenced by the eating habits of not only the mother but also the father. Obesity in men and women has consequences for offspring through nutritional programming. Nutritional programming of the offspring occurs through changes in lipid metabolism and leads to an increased risk for cardiometabolic diseases (101).

Metabolic changes in pregnant women occur in early pregnancy. In the first weeks after conception, the placenta begins to secrete hormones that affect the metabolism of all nutrients. Human chorionic gonadotropin (hCG) is detected in serum and urine a few days after implantation and is produced by trophoblast cells that surround the growing embryo (initially syncytiotrophoblast), which forms the placenta after implantation. Its concentrations increase during early pregnancy and reach a maximum during the first 2 months after conception, after which they decrease, maintaining low values until the date of delivery. Its role is immunomodulatory, reflected in the attraction of Treg in the uterus and the mother's tolerance toward fetal recognition (102).

The balance between Treg and Th17 cells is necessary to maintain the normal course of pregnancy, while a change in the ratio of Th17/Treg and Th17 cells is a potential cause of pregnancy complications, including preeclampsia, recurrent spontaneous abortion, and gestational diabetes mellitus (103). Decidual immune cells not only regulate the maternal immune response to promote fetal hemi-allograft tolerance but also mediate trophoblast implantation and invasion. At the same time, trophoblasts mediate fetal–maternal interactions for the exchange of nutrients, gases, and waste products, as well as for the regulation of immune tolerance (104). Human placental lactogen (hPL) is a polypeptide hormone produced by syncytiotrophoblast cells of the placenta. hPL increases progressively during pregnancy, and its full function has not been fully elucidated (105).

Because it is biologically similar to growth hormone, it is possible that it represents a type of growth factor for the fetus and placenta. In addition, it has a significant metabolic role in the metabolism of carbohydrates and lipids, as well as the availability of nutrients for the fetus. The placenta is the primary source of steroid hormones, including the synthesis of estrogen and progesterone (106). Estrogens and progesterone, in addition to affecting the reproductive organs, play a role in the regulation of insulin and glucose homeostasis, lipid metabolism, and appetite regulation, which can be important in promoting metabolic changes in the mother during pregnancy. In addition, progesterone and estrogen have opposite effects on food intake. Estrogen decreases food intake in part by inducing leptin production in adipose tissue, while progesterone increases food intake by increasing neuropeptide I (NPI) levels and decreasing cocaine peptide and amphetamine-regulated transcript (CART) expression from sides of the hypothalamus (107). The course of early pregnancy can be called the anabolic phase, while the course of late pregnancy is accompanied by complex metabolic processes thanks to the interaction of mother and fetus and is designated as the catabolic phase (108).

Fatty acids participate in the synthesis of cell membranes and are also a source of energy. During the first weeks of pregnancy, cholesterol is transferred to the placenta with the help of LDL cholesterol, from where it is later released into the fetal circulation. Cholesterol participates in the synthesis of steroid hormones and cell membranes. During the first weeks of pregnancy, serum triglycerides are 20% higher, and by the end of pregnancy they reach three times higher concentrations compared to nonpregnant women. Other serum lipids (cholesterol, phospholipids, fatty acids) also increase during pregnancy but to a lesser extent. Fat accumulation occurs as a result of increased sensitivity to insulin and an increased number of insulin receptors in adipose tissue, and thus increased activity of tissue lipoprotein lipase (109).

Lipid metabolism is an important factor in fetal development, course, and outcome of pregnancy. In healthy pregnant women, triglyceride values increase two to four times, while the increase in cholesterol is around 25–50%. However, the mechanism leading to these adaptive changes has not been fully elucidated (110).

Pregnancy is accompanied by an increase in LDL cholesterol and free fatty acids. The value of HDL cholesterol increases in the second trimester and maintains the achieved value until the end of pregnancy. These changes during pregnancy represent a necessary adaptation of the pregnant woman's metabolism to the demands of the fetus and prepare the pregnant woman for the upcoming birth and lactation (111).

An increase in the value of triglycerides in the plasma corresponds to an increase in the value of VLDL lipoprotein particles. Via cholesteryl ester transfer protein, whose activity increases in mid-pregnancy, there is an exchange of triglycerides between VLDL and LDL or HDL particles, so there is an increased accumulation of triglycerides in these particles as well (112).

During pregnancy, hepatic lipase (HL) activity declines, allowing the conversion of HDL-2 particles to HDL-3 lipoprotein particles, leading to the accumulation of triglyceride-rich HDL-2 particles (113). The physiological increase in lipids plays an essential role during pregnancy, which is reflected in the mobilization of free fatty acids from fat depots in late pregnancy, which are necessary for fetal growth and steroid synthesis in placental tissue (114, 115).

Free fatty acids are necessary for the development of the fetal brain and visual system, and as biosynthetic precursors for hormones, they pass through the placenta by oxal diffusion (116).

Fatty acids are first converted to acyl-CoA. Acyl-CoA can then be esterified into various lipids, triglycerides, phospholipids, and cholesterol esters. For the passage of fatty acids to the fetus, the role of the placenta is played by the expression of genes for placental lipase enzymes, which include LPL, endothelial lipase, and hormone-sensitive lipase, which release fatty acids from acyl-CoA, triglycerides, and phospholipids (117).

In addition, the placenta has a large number of receptors for binding VLDL particles. In the catabolic phase, placental hormones increase the production of VLDL particles and decrease the activity of hepatic lipase, which, together with the increased expression of the VLDL apoE receptor in the placenta, enables the appropriate influx of maternal triglyceride-rich lipoproteins into the fetoplacental unit (118).

The cause of intrauterine growth of the fetus can be an insufficient supply of lipids to the fetus, which is associated with disturbed fetoplacental flow associated with an abnormal lipid profile, which is reflected in reduced values of total cholesterol, triglycerides, LDL, and HDL-cholesterol retardation (119).

LDL cholesterol contains seven subfractions, that is, particles that differ in size, density, chemical composition, and function in the body. LDL1–2 is a subfraction of larger particles that are antiatherogenic, while a fraction of smaller and denser LDL 3–7 particles, more sensitive to oxidative modifications compared to normal-density LDL particles, shows greater binding to vessel wall proteoglycans and reduced LDL receptor uptake (9).

Highly atherogenic small dense lipoproteins LDL3–7 (sdLDL) dominate in 97% of women in the 35th and 36th weeks of pregnancy. Small dense LDL3–7 particles are thought to contribute to endothelial dysfunction in preeclampsia (120).

Considering the specifically changed lipid profile during pregnancy, the question arises whether hyperlipidemia in pregnancy is dyslipidemia. Classic dyslipidemia is characterized by a decrease in HDL cholesterol, which is not the case during an uncomplicated pregnancy. In preeclampsia, this characteristic of classic dyslipidemia with a decrease in HDL cholesterol is seen (121).

Triglycerides are esters of glycerol and fatty acids. Esterification is carried out with one, two, or all three hydroxyl groups of glycerol with fatty acids, whereby mono-, di-, or triglycerides are formed. Their basic role in the body consists in the creation of energy depots from which fatty acids are released as needed, and their oxidation provides the energy necessary for the life of all cells. The largest amount of triglycerides is found in the composition of fatty tissue (about 95%), while small amounts are present in the blood. Even for triglycerides, it has now been reliably established that their increased amount in the blood plays a significant role in the process of atherosclerosis (122).

During pregnancy, the TG/HDL-chole index is particularly interesting, considering the manifest hypertriglyceridemia during pregnancy. High TG can be complicated by low HDL-C. Although both parameters are mutually independent risk factors for ASCVD, the most relevant cases are those with simultaneously high TG and low HDL-C, which has a strong atherogenic effect. It has been proven that a ratio (TG/HDL-chole) higher than 0.5 is considered an indicator of atherogenicity (123, 124). In pregnant women, VLDL levels increase due to decreased activity of LPL and hepatic lipase. During pregnancy, VLDL production is also facilitated by hormone-sensitive lipase in adipose tissue (125).

It is considered that the increase in the value of VLDL particles during pregnancy is due to the increased synthesis of these particles, and not to compositional changes within the particles, considering that the VLDL/Chol index does not change significantly during pregnancy and is lower in all trimesters of pregnancy compared to nonpregnant women.

Therefore, although the lipid metabolism in pregnancy is particularly changed, taking into account lipid and lipoprotein indices, it is a balanced hyperlipidemia (126).

Studies have shown that there is a difference between triglycerides and LDL subfractions of the "threshold" effect. Namely, with low and normal triglyceride values, there is a positive correlation with the LDL-2 subfraction, and with triglyceride values above 1.5 mmol/L, there is a positive correlation with the LDL-3 subfraction, which is relatively constant below the above triglyceride values. In the first trimester of pregnancy, increased triglycerides may be responsible for the increase in small dense LDL particles (127).

The total cholesterol in the fetal plasma increases proportionally with the cholesterol values of the pregnant woman during the second trimester of pregnancy and slowly decreases with the progress of the pregnancy, that is, the gestational age of the fetus. It has been proven that the concentrations of lipids in the umbilical cord blood of fetuses of healthy pregnant women are significantly lower than the lipids in the blood of pregnant women, and elevated HDL cholesterol and a lower ratio of LDL-chole/HDL-chole in umbilical cord blood indicate that the fetus is protected from atherogenic lipids during a healthy pregnancy. On the other hand, it is believed that hypercholesterolemia during pregnancy, although limited and transient, can cause the development of pathological changes and fatty streaks in the aorta of the fetus, thereby affecting the development of atherosclerosis in the later life of the offspring (128).

The reason for fetal fat accumulation may be gender differences despite similar placental exposure to excess TG or FFA. Total body fat at birth is higher in female infants from 38 to 42 weeks of gestation compared to the amount of body fat in male infants. The reason for gender differences in the amount of fat in newborns is the reaction of the placenta to the same environment depending on the sex of the fetus.

Epigenetic mechanisms that regulate placental differentiation and function, and thus fetal growth and phenotype, may have contributed to this programming process. The metabolic status of the mother can change the expression and function of the placental gene with direct consequences on the development of the fetal tissue, i.e., the influence on the phenotype of the offspring (129), primarily due to the effect of estrogen. The antiatherogenic effect of HDL cholesterol is explained by its functional protection, through the transport of cholesterol from peripheral tissues to the liver, and also by its anti-inflammatory and antioxidant role. When it comes to HDL cholesterol subfractions, we usually talk about two fractions: HDL2, which is larger, and HDL-3, which is smaller and denser and has a lower atheroprotective character (130).

It is believed that HDL is also susceptible to lipid peroxidation, even more easily than LDL, and because it is a smaller molecule than LDL, it diffuses more easily into the subendothelial space. When it comes to HDL subfractions, the HDL-2 fraction is more susceptible to oxidation than the HDL-3 fraction (131).

During pregnancy, numerous adaptive mechanisms of changes in glucose and lipid metabolism occur to meet the growth and development needs of the fetus. Abnormal glucose and lipid metabolism are associated with fetal birth weight in the second and third trimesters of pregnancy (132). A high birth weight of a newborn leads to increased short-term complications for the newborn and the pregnant woman, and in the long term, metabolic diseases such as obesity and type 2 diabetes can occur with a significantly increased risk of cardiovascular diseases. The theory of the fetal origin of adult diseases supports a series of changes that affect the fetus during the intrauterine period, which is reflected in the physical condition of the child and adult (133).

Maternal triglycerides and free fatty acids may be potential substrates for fetal fat development and storage. Women enter pregnancy with pre existing overweight/obesity, features of metabolic syndrome, and poor dietary patterns. When insulin resistance (IR) builds up during pregnancy, the placenta is exposed to excess nutrients that are broken down by placental lipases, metabolized or transported, primarily glucose and amino acids, to the fetus. High maternal glucose and amino acid levels promote fetal hyperinsulinemia, leading to increased fetal subcutaneous fat. High exposure to free fatty acids (FFA) may promote increased fetal intrahepatic lipid storage, serving as one of the multiple 'first hits' that lead to childhood non-alcoholic fatty liver disease (NAFLD) and propagate increased birth weight. Maternal obesity further alters the differentiation potential of fetal umbilical cord mesenchymal stem cells, increasing their propensity to differentiate into adipocytes versus skeletal muscle cells, which also contributes to greater adiposity at birth. The combination of maternal overnutrition as a function of pre- and pregnancy-IR with poor lifestyle, exposure of the fetus and placenta to excess nutrients, and fetal hyperinsulinemia influences increased adiposity at birth emphasizing increased risk for later childhood obesity, NAFLD, and metabolic diseases (134).

Interest in high-density lipoproteins (HDL) has focused on their possible protective role in atherosclerotic cardiovascular disease (ASCVD). Through evolution from invertebrates, HDLs are the main lipoproteins, while lipoproteins containing apolipoprotein B first developed in vertebrates. HDLs transport cholesterol and other lipids between different cells like a reusable ferry, but they serve many other functions including communicating with cells and inactivating biological hazards like bacterial lipopolysaccharides. Recent studies indicate that functional measures of HDL, particularly cholesterol efflux capacity, HDL particle number, or different HDL proteins, are better predictors of ASCVD events than HDL-C. Low HDL-C is observationally, as well as genetically, associated with an increased risk of infectious diseases, death during sepsis, diabetes mellitus, and chronic kidney disease. Additional, but only observational, data indicate an association of low HDL-C with various autoimmune diseases and cancer, as well as all-cause mortality. Conversely, extremely high levels of HDL-C are associated with an increased risk of age-related macular degeneration (also genetic), infectious diseases, and all-cause mortality (135).

Inflammation is a key driver of chronic diseases such as ASCVD and diabetes, as well as infection and malignancy. HDL directly affects the inflammatory process and, conversely, inflammation affects the function of HDL. Approximately ~90-95% of plasma apoA-I is bound to HDL particles. While early studies focused on the anti-inflammatory effects of HDL, recent studies showing that HDL and apoA-I may also be pro-inflammatory may be one explanation for the limited success of HDL-raising drugs in reducing ASCVD. However, HDL's pro-inflammatory effects may be beneficial in other diseases, such as sepsis, where increased inflammation can promote efficient bacterial clearance. Understanding the mechanistic links between HDL and its role in inflammation is vital to understanding the prognostic and therapeutic potential of this class of lipoproteins (136).

An altered lipid profile during pregnancy has the characteristics of a proatherogenic lipid profile and may influence the development of cardiovascular disease (CVD) in a woman's later life. Certain studies have shown a positive correlation between the number of pregnancies and the risk of developing CVD in later life, which may be explained by lower HDL cholesterol levels in later life (137). Also, studies have shown that women with more pregnancies have a higher risk of developing diabetes mellitus later in life (138). Breastfeeding also affects lipid metabolism. Human milk (HM) contains a large number of lipids that are necessary for optimal growth and development. These lipids make up approximately 5% of the total milk profile and are packaged as globules of milk fat.

Milk fat globules are synthesized from a triacylglycerol droplet of the inner endoplasmic reticulum of the milk fat epithelium, so the milk fat globule membrane is a three-layer structure composed of many polar lipid circles located in a circle. Breast milk fat is the main source of energy, providing over 40% of the total energy for babies during the first 6 months of life. Some of the essential lipid components include PUFAs and long-chain PUFAs (LC-PUFAs) of the *n*-6 and *n*-3 series and lipid-soluble vitamins. The latter study reported complex interactions between fatty acid and protein profiles affected by lactation and stadium gestation. Regional differences in diet may also affect the composition of the highest milk; for example, the increased consumption of vegetable oil products containing *n*-3 PUFA (linoleic acid) in the Western diet has led to 50% to 90% PUFA (139).

The effect of the largest milk on the initial gut microbiota also prevents the expression of immune-induced diseases (asthma, inflammatory bowel disease, and type 1 diabetes) later in life through a balanced initial immune response to the food source (140).

6.6.1 Oxidative Stress in Pregnancy

Pregnancy is a state of increased sensitivity to oxidative stress, which occurs mainly due to the increased need for oxygen in the mitochondria of the placenta. A full-term pregnancy involves a series of events including embryogenesis, embryo implantation, fetoplacental development, fetal growth, and delivery. Pregnancy increases oxidative stress, a phenomenon arising from the normal systemic inflammatory response, resulting in large amounts of circulating reactive oxygen species (ROS) produced by the placenta (141).

Oxidative stress during pregnancy can lead to cell damage. In situations where oxidative stress overcomes the antioxidant defense of the placenta, tissue damage occurs (142). Oxidative stress represents a disturbance in the balance of oxidation-reduction processes in the body that occurs due to excessive production of oxygen free radicals, which cellular homeostatic mechanisms are unable to neutralize. Oxygen free radicals play an important role as secondary messengers in many cellular signaling processes, and their high reactivity originates from one or more unpaired electrons. When the balance with the antioxidant system is disturbed, which happens when oxygen free radicals are in high concentrations, there is serious damage to biological molecules, which results in a series of pathophysiological and pathological changes, including cell death. Therefore, oxidative stress plays a significant role in the pathogenesis of many complications that may occur during pregnancy and may affect fetal development (143).

The high reactivity of molecular oxygen enables its participation in high-energy processes that take place in cells containing mitochondria, where the process of oxidative phosphorylation takes place (144).

Over 90% of oxygen from the air in the body is reduced to water by receiving four electrons from the electron transport system in the mitochondrial respiratory chain (145). Molecular oxygen in this system is used to obtain water, and the successively released energy during electron transport is used to obtain electrochemical potential and ATP synthesis. Electron transport is a stepwise process that begins with hydrogen, i.e., coenzyme dehydrogenase, and ends with the cytochrome c oxidase enzyme complex (146). The mitochondrial electron transport system is one of the most important processes in which the superoxide anion radical is generated. Molecular oxygen, diffusely present in cells, is tightly bound to the cytochrome c oxidase enzyme complex (147, 148).

Superoxide anion can also be formed within the shorter electron transport chain within the endoplasmic reticulum during protein synthesis and biotransformation of exogenous and endogenous substrates (149). Other sources of superoxide under physiological conditions include NADPH oxidase, cytochrome P450, and other oxidoreductases. Various growth factors, drugs, toxins, inflammation, ultraviolet exposure, and lipid peroxidation can also lead to increased ROS generation. The enzyme xanthine dehydrogenase under normal conditions breaks down purines, xanthine, and hypoxanthine to uric acid, using NAD^+ as an electron acceptor. In hypoxic conditions, the formation of hypoxanthine increases with the transfer of electrons to molecular oxygen and the formation of superoxide anions (150).

Superoxide is neutralized in the presence of the enzyme superoxide dismutase (SOD), producing hydrogen peroxide (H_2O_2), which is less reactive and is not a free radical, but since it participates in the creation and detoxification of free radicals, it is considered part of ROS. Hydrogen peroxide is degraded to water by catalase and glutathione peroxidase and is considered an oxygen metabolite in aerobic cellular metabolism (151).

Hydrogen peroxide occurs in normal mammalian cell metabolism and is an important metabolite in oxidative stress. The term *oxidative eustress* refers to physiological oxidative stress as opposed to overload, *oxidative distress*, which causes oxidative damage to cells (152).

Numerous factors in physiology can cause oxidative stress. The most common factors are the external environment, air pollution, inadequate nutrition, ionizing radiation, the use of antibiotics, as well as diseases that occur during pregnancy. In the last 3 years, the most common factor of oxidative stress during pregnancy has been infection with the SARS-CoV-2, the virus causing the COVID-19 disease. Respiratory distress syndrome in a pregnant woman can affect the supply of oxygen to the fetus and trigger the mechanism of metabolic disorders of the fetus and newborn caused by asphyxia. In the absence of oxygen in utero in the placenta, a microcirculation disorder occurs, which leads to a metabolic disorder in the mother–fetus system because there is a disorder in the exchange of nutrients, and later also to a disorder in oxygen transport (153). Insufficient oxygenation leads to inadequate maturation of surfactant, premature birth, accelerated breathing of the newborn, and the development of pneumothorax. Neonates born by cesarean section (CS) often have "wet lungs" accompanied by forced breathing that can lead to a pneumothorax (154). In infants born via CS at term or moderately preterm (30–36 weeks), the incidence of NP and respiratory problems is significantly increased (155).

6.6.2 Consequences of the Action of Free Oxygen Radicals During Pregnancy

Lipid peroxidation, resulting from the action of free radicals on the polyunsaturated fatty acids of lipids, is present in all cells and tissues. Since lipids are an integral part of membranes, numerous cell organelles that contain membranes are exposed to damage caused by uncontrolled lipid peroxidation, which jeopardizes not only the functioning of the cell but also its survival (156).

During the process of lipid peroxidation, the fluidity and permeability of the cell membrane change, the transport of electrolytes through the membrane is disturbed, the electrical resistance decreases, and the mobility of membrane proteins occurs. Lipid peroxidation is considered the cause of numerous diseases, such as atherosclerosis, diabetes, Parkinson's disease, Alzheimer's disease, and various chronic inflammatory conditions. It is also considered a cause of complications in pregnancy and is particularly crucial in the development of preeclampsia (157).

Uncontrolled lipid peroxidation also leads to pathological changes during pregnancy in both mother and fetus. The placenta plays a central role in fetal development, growth, mother–fetus communication, homeostasis, and adaptation of pregnancy to injury, so the placenta plays an important role in pregnancy complications, such as fetal growth restriction, preeclampsia, preterm labor, and abruption (158).

The mechanisms by which ferroptosis can play a role in placental dysfunction are as follows: the placenta is normally susceptible to hypoxia-reoxygenation in early pregnancy, as well as before and during labor under the influence of uterine contractions; iron depots in placental trophoblasts are actively transferred across the placenta to the fetus in development; lower levels of glutathione peroxidase 4 (GPX4), a key enzyme that protects cells from the accumulation of harmful specific hydroxy-peroxidized phospholipids (Hp-PL) and ferroptosis. All these are associated with trophoblast damage, placental dysfunction, and the onset of preeclampsia (159).

Although lipid peroxidation had drawn the bulk of researchers' attention, recently protein oxidation, due to its large presence and numerous functions in the body, has been recognized as a possible main target of free radical action. The side chains of all amino acid residues of proteins are sensitive to oxidation caused by the action of ROS. Lipid peroxidation is associated with protein aggregation and leads to disruption of organelles and cellular functions. Maintaining the redox status of proteins is thought to be fundamental to maintaining normal cell function. Protein oxidation, and thus altered structure and function, as well as disturbed homeostasis of plasma proteins, are associated with neurodegenerative diseases (160, 161).

Highly reactive oxygen and nitrogen species (RONS) carry out lipid peroxidation of the cell membrane leading to intracellular accumulation of highly reactive molecules such as lipid peroxide, malondialdehyde, 4-hydroxynonenal, and acrolein, which change the structure and thus the function of proteins (162). Unsaturated fatty acids such as docosahexaenoic acid, eicosapentaenoic acid, and arachidonic acid are the most sensitive to the influence of RONS. Antioxidants can modulate lipid peroxidation in two ways, by directly scavenging RONS or indirectly by increasing and decreasing lipoxygenase enzymes (163, 164).

In a study of total cholesterol concentration on the second day of life (77), lower values of total cholesterol were found in the group of newborns with hypoxia on oxygen therapy, which can be explained by the peroxidation of unsaturated phospholipids and cholesterol in cell membranes and lipoproteins caused by oxidative stress that occurs when the amount of prooxidants exceeds the antioxidant capacity by forming ROS such as superoxide (O^{2-}), hydrogen peroxide (H_2O_2), and hydroxyl radical (HO), which can initiate non-enzymatic peroxidation of lipids mediated by free radicals (165).

The propensity of lipoproteins toward lipid peroxidation depends on the type and amount of lipids – that is, the type and amount of apolipoproteins. LDL contain exclusively Apo B-100 and are mostly oxidized lipids. This susceptibility to oxidation can be attributed to the properties of the membrane and the outer surface of the nucleus, where the two lipid envelopes are located. The cell membrane is most susceptible to peroxidation due to the high content of unsaturated fatty acids, leading to a lower concentration of LDL cholesterol on the second day of life (166).

HDL have the highest percentage of Apo A-I in their composition, which enables the esterification of free cholesterol on the surface of the particle. Apo A-II is represented in a slightly smaller percentage. Lipid peroxidation affects lipids, cholesterol, and apolipoprotein HDL particles, reducing the concentration of HDL cholesterol in the plasma on the second day of life. In experiments on animals, it has been proven that oxidized HDL particles have a protective effect on the vascular wall and on cells that are susceptible to lipid peroxidation processes (167).

Triglycerides are esters of glycerol and fatty acids that are part of the lipid part of lipoprotein particles. During hypoxia, there is lipid oxidation of lipoprotein particles, breakdown of fatty acids, formation of sphingosine-1-phosphate, and Ca^{2+}-mediated apoptosis, which act chemotaxically on macrophages leading to phagocytosis of cells or cell fragments (168, 169).

Reduced concentrations of triglycerides in plasma on the second day of life occur as a result of the action of lecithin cholesterol-acyltransferase, which in conditions of intrapartum acidosis leads to the oxidation of fatty acids of fatty tissue triglycerides (170).

In the plasma membrane, where most of the cellular cholesterol is located, peroxidation leads to structural and functional damage to the membrane. Iron-stimulated peroxidation of cholesterol and other lipids leads to a new form of regulated cell death called ferroptosis. Cholesterol as well as other unsaturated lipids are susceptible to peroxidative degradation after exposure to strong antioxidants such as hydroxyl radical or peroxynitrite formed under conditions of oxidative stress. Cholesterol peroxidation leads to the breakdown of fatty acid residues and the release of malondialdehyde and other products, as well as lysolecithin, which is a chemotaxis agent for monocytes and T lymphocytes. Monocytes "take" lipoproteins and gradually turn into foam cells. There is activation of complement and an inflammatory reaction on the intima of the blood vessel, which, by creating atherogenic plaques, can become a trigger for numerous diseases of the organs in which the process occurred during early neonatal life (165, 171).

6.7 Conclusion

Deoxyribonucleic acid (DNA) damage is considered the most serious negative response of the cell to the excessive production of ROS and the oxidation of these molecules because they can cause mutations and genetic instability that is reflected in the outcome of pregnancy itself, whether it is complications during pregnancy or the quality and health of the fetus and newborn. Although the DNA repair system is very efficient, excessive production of ROS leading to mitochondrial damage activates an inflammatory process that activates lymphocytes and macrophages, creating an inflammatory process with damage at the level of purine and pyrimidine bases. DNA oxidation is considered a significant factor in the development of carcinogenesis, diabetes, and aging.

References

1. Quiney JR,Wats GF. (eds). *Classic papers in hyperlipidaemia*. London: MSD Science Press;1989, pp. 25–31.
2. Goldstein JL, Brown MS. The LDL receptor. *Arterioscler Thromb Vasc Biol*. 2009;29:431–438.
3. Sniderman AD, Cianfore K. Measurement of apolipoproteins: time to improve the diagnosis and treatment of the atherogenic dyslipoproteinemias. *Clin Chem*. 1996;42:489–491.
4. Ikeda Y, Ashida Y, Takagy A. et al. Mechanism of the production of small dense LDL(s LDL) in hypertriglyceridemia. In: Jacotot B, Mathe D, Fruchart J-C, editors. *Atherosclerosis XI*. Amsterdam- Lausanne-New York-Oxoford-Singapore-Tokyo: Elsevier; 1998, pp. 777–788.
5. Jovandaric Z Miljana, Milenkovic J Svetlana. Significance of lipid and lipoprotein in organism. In: Waisundara VZ, Jovandaric M, editors. *Apolipoproteins, triglycerides and cholesterol*. London, UK: IntechOpen; 2020, pp. 3–11.
6. Goldstein JL, Brown MS. Molecular medicine: *the cholesterol quartet. Science*. 2001;292:1310–1312.
7. Khoury S, Canlet C, Lacroix MZ, Berdeaux O, Jouhet J, Bertrand-Michel J. Quantification of lipids: model, reality, and compromise. *Biomolecules*. 2018;8:174.
8. Duttaroy AK. Transport of fatty acids across the human placenta: a review. *Prog Lipid Res*. 2008;48:52–61.
9. Jovandaric MZ, Ivanovski PI. Free fatty acids of newborns from women with gestational diabetes mellitus. *Fetal Pediatr Pathol*. 2017;36:190–194.

10. Luo J, Yang H, Song BL. Mechanisms and regulation of choles terol homeostasis. *Nat Rev Mol Cell Biol.* 2020;21:225–245.
11. Luo J, Jiang LY, Yang H, Song BL. Intracellular cholesterol transport by sterol transfer proteins at membrane contact sites. *Trends Biochem Sci.* 2019;44:273–292.
12. Paalvast Y, de Boer JF, Groen AK. Developments in intestinal cholesterol transport and triglyceride absorption. *Curr Opin Lipidol.* 2017;28:248–254.
13. Grabner GF, Xie H, Schweiger M, Zechner R. Lipolysis: cellular mechanisms for lipid mobilization from fat stores. *Nat Metab.* 2021;3:1445–1465.
14. Hurt-Camejo E, Pedrelli M. Why are brown bears protected against atherosclerosis even though their plasma cholesterol levels are twice that of humans? *Clin Investig Arterioscler.* 2022;34:322–325.
15. Hatch GM. Cell biology of cardiac mitochondrial phospholipids. *Biochem Cell Biol.* 2004;82:99–112.
16. Vance JE, Vance DE. The assembly of lipids into lipoproteins during secretion. *Experientia.* 1990;46:560–569.
17. Hirano T. Pathophysiology of diabetic dyslipidemia. *J Atheroscler Thromb.* 2018;25:771–782.
18. Jovandaric MZ, Babic S, Raus M, Medjo B. The importance of metabolic and environmental factors in the occurrence of oxidative stress during pregnancy. *Int J Mol Sci.* 2023;24:11964.
19. Zhang X, Lei D, Zhang L, Rames M, Zhang S. A model of lipid-free apolipoprotein A-I revealed by iterative molecular dynamics simulation. *PLoS One.* 2015;10:e0120233.
20. Segrest JP, Li L, Anantharamaiah GM, Harvey SC, Liadaki KN, Zannis V. Structure and function of apolipoprotein A-I and high-density lipoprotein. *Curr Opin Lipidol.* 2000;11:105–115.
21. Bassett GR, Gillard BK, Pownall HJ. Cholesterol determines and limits rHDL formation from human plasma apolipoprotein A-II and phospholipid membranes. *Biochemistry.* 2012;51:8627–8635.
22. Jiang ZY, Han TQ, Suo GJ, Feng DX, Chen S, Cai XX, Jiang ZH, Shang J, Zhang Y, Jiang Y, Zhang SD. Polymorphisms at cholesterol 7alpha-hydroxylase, apolipoproteins B and E and low density lipoprotein receptor genes in patients with gallbladder stone disease. *World J Gastroenterol.* 2004;10:1508–1512.

23. Drouin-Chartier JP, Tremblay AJ, Lépine MC, Lemelin V, Lamarche B, Couture P. Substitution of dietary ω-6 polyunsaturated fatty acids for saturated fatty acids decreases LDL apolipoprotein B-100 production rate in men with dyslipidemia associated with insulin resistance: a randomized controlled trial. *Am J Clin Nutr.* 2018;107:26–34.
24. Hanson CL, Ilag LL, Malo J, Hatters DM, Howlett GJ, Robinson CV. Phospholipid complexation and association with apolipoprotein C-II: insights from mass spectrometry. *Biophys J.* 2003;85:3802–3812.
25. Morton AM, Koch M, Mendivil CO, Furtado JD, Tjønneland A, Overvad K, Wang L, Jensen MK, Sacks FM. Apolipoproteins E and CIII interact to regulate HDL metabolism and coronary heart disease risk. *JCI Insight.* 2018;3:e98045.
26. Guardamagna O, Abello F, Anfossi G, Pirro M. Lipoprotein(a) and family history of cardiovascular disease in children with familial dyslipidemias. *J Pediatr.* 2011;159:314–319.
27. Haberkant P, Holthuis JCM. Fat & fabulous: bifunctional lipids in the spotlight. *Biochim Biophys Acta.* 2014;1841:1022–1030.
28. Thiam AR, Farese RV Jr, Walther TC. The biophysics and cell biology of lipid droplets. *Nat Rev Mol Cell Biol.* 2013;14:775–786.
29. McDonough PM, Ingermanson RS, Loy PA, Koon ED, Whittaker R, Laris CA, Hilton JM, Nicoll JB, Buehrer BM, Price JH. Quantification of hormone sensitive lipase phosphorylation and colocalization with lipid droplets in murine 3T3L1 and human subcutaneous adipocytes via automated digital microscopy and high-content analysis. *Assay Drug Dev Technol.* 2011;9:262–280.
30. Gimpl G. Cholesterol-protein interaction: methods and cholesterol reporter molecules. *Sub-cell Biochem.* 2010;51:1–45.
31. Redgrave TG. Formation of cholesteryl ester-rich particulate lipid during metabolism of chylomicrons. *J Clin Invest.* 1970;49:465–471.
32. Wong H, Davies RC, Thuren T., et al. Lipoprotein lipoze domain function. *J Biol Chem.* 1994;269:10319–10323.
33. Brown MS, Goldstein JL. A receptor-mediated pathway for cholesterol homeostasis. *Science.* 1986;232:34–36.

34. Krieger M, Herz J. Structure and function of multiligand lipoprotein receptors: macrophage scavenger receptors and LDL receptor-related proteins. *Ann Rev Biochem.* 1994;63:601–637.
35. März W, Scharnagl H, Kleber M, Silbernagel G, Nauck M, Müller-Wieland D, von Eckardstein A. Labordiagnostik von Fettstoffwechselstörungen [Laboratory diagnostics of lipid metabolism disorders]. *Dtsch Med Wochenschr.* 2023;148:e120–e146.
36. Brown MS, Goldstein JL. A receptor mediated patway for cholesterol homeostasis. *Science.* 1986;232–234.
37. Brunzell JD. Familial lipoprotein lipase deficiecy and other cause of the chylomicronemia syndrome. In: Scriver CR et al., editors. *The metabolic basis of inherited disease*, 6th ed. New York: McGraw-Hill; 1989, p. 45.
38. Goldstein JL, Brown MS. Familial hyperholesterolemia. In: Scriver CR et al., editors. *The metabolic basis of international disease*, 6th ed. New York:McGraw-Hil; 1989, p. 45.
39. Orchard T. Dyslipoproteinaemia and diabetes. *Endocrinol Metab Clin N Am.* 1990;2:361.
40. Linton MF, Tao H, Linton EF, Yancey PG. SR-BI: a multifunctional receptor in cholesterol homeostasis and atherosclerosis. *Trends Endocrinol Metab.* 2017;28:461–472.
41. Buonocore G, Bracci R, Perrone S, Vento M. Oxygen toxicity. In: Buonocore G, Bracci R, Weindling M, editors. *Neonatology a practical approach to neonatal diseases.* Milano, Italy: Springer-Verlag; 2012, pp. 242–249.
42. Mirończuk-Chodakowska I, Witkowska AM, Zujko ME. Endogenous non-enzymatic antioxidants in the human body. *Adv Med Sci.* 2017;63:68–78.
43. Lushchak VI. Free radicals, reactive oxygen species, oxidative stress and its classification. *Chem.-Biol. Interact.* 2014;224:164–175.
44. Ye ZW, Zhang J, Townsend DM, Tew KD. Oxidative stress, redox regulation and diseases of cellular differentiation. *Biochim Biophys Acta.* 2015;1850:1607–1621.
45. Lefaki M, Chondrogianni N. Redox regulation of proteasome function. *Redox Biol.* 2017;13:452–458.

46. Delivoria-Papadopoulos M. Mechanisms of cerebral injury in perinatal asphyxia and stragies for prevention. *J Pediatr.* 1998;132:30–34.
47. Chan PH, Fishman RA. Transient formation of superoxide radicals in polyunsaturated fatty acids-induced brain swelling. *J Neurochem.* 1980;35:1004.
48. Tsukahara H. Biomarkers for oxidative stress: clinical application in pediatric medicine. *Curr Med Chem.* 2007;14:339–351.
49. Saugstad OD, Sejersted Y, Solberg R,Wollen EJ, Bjørås. Oxygenation of the newborn: a molecular approach. *Neonatology.* 2012;101:315–325.
50. Jovandaric MZ, Milenkovic SJ. Neurological impairments in hypoxic neonates and lactate levels. *Neurol Res.* 2018;40:822–827.
51. Puchalska P, Crawford PA. Multi-dimensional roles of ketone bodies in fuel metabolism, signaling, and therapeutics. *Cell Metab.* 2017;25:262–284.
52. Puchalska P, Crawford PA. Metabolic and signaling roles of ketone bodies in health and disease. *Annu Rev Nutr.* 2021;41:49–77.
53. Rajendran M, Dane E, Conley J, Tantama M. Imaging adenosine triphosphate (ATP). *Biol Bull.* 2016;231:73–84.
54. Okamoto K, Kuwabara Y, Nishino T. Molecular switching mechanism of xanthine oxidoreductase. *Tanpakushitsu Kakusan Koso.* 2004;49:625–633.
55. Stipek S, Novak L, Crkovska J, Zima T, Platenik J. Xanthine oxidoreductase: biochemical, biological and pathogenic functions. *Sb Lek.* 1994;95:289–295.
56. Parks DA, Granger DN. Xanthine oxidase: biochemistry, distribution and physiology. *Acta Physiol Scand Suppl.* 1986;548:87–99.
57. Farber JL. Mechanisms of cell injury by activated oxygen species. *Environ Health Perspect.* 1994;102 Suppl 10:17–24.
58. Harned J, Nagar S, McGahan MC. Hypoxia controls iron metabolism and glutamate secretion in retinal pigmented epithelial cells. *Biochim Biophys Acta.* 2014;1840:3138–3144.
59. Xu KY, Li F. Iron metabolism and neonatal hypoxic ischemic brain damage. *Sheng Li Xue Bao.* 2017;69:218–224.
60. Mihajlović Lj, Mihajlović N, Petrović A, Pavlović R. Molekularni mehanizam ćelijskih oštećenja pri akutnoj hipoksiji. *Acta Medica Medianae.* 2002;41:51–61.

61. Rothman SM, Olney JW. Glutamate and the pathophysiology of hypoxic—ischemic brain damage. *Ann Neurol.* 1986;19:105–111.
62. Mayer ML. The challenge of interpreting glutamate-receptor ion-channel structures. *Biophys J.* 2017;113:2143–2151.
63. Rousseaux CG. A review of glutamate receptors II: pathophysiology and pathology. *J Toxicol Pathol.* 2008;21:133–173.
64. Scorrano L, Oakes SA, Opferman JT, Cheng EH, Sorcinelli MD, Pozzan T, Korsmeyer SJ. BAX and BAK regulation of endoplasmic reticulum Ca^{2+}: a control point for apoptosis. *Science.* 2003;300:135–139.
65. Rousset CI, Baburamani AA, Thornton C, Hagberg H. Mitochondria and perinatal brain injury. *J Matern Fetal Neonatal Med.* 2012;25 Suppl 1:35–38.
66. Graham SH, Chen J. Programmed cell death in cerebral ischemia. *J Cereb Blood Flow Metab.* 2001;21:99–109.
67. Thornton C, Hagberg H. Role of mitochondria in apoptotic and necroptotic cell death in the developing brain. *Clin Chim Acta.* 2015;451(Pt A):35–38.
68. Lockshin RA, Zakeri Z. Caspase-independent cell deaths. *Curr Opin Cell Biol.* 2002;14:727–733.
69. Abe K, Aoki M, Kawagoe J, Yosida T, Hattori A, Kogure K, Itoyan Y. Ischemic delayed neuronal death, a mitochondrial hypothesis. *Stroke.* 1995;26:1478–1489.
70. Torres-Cuevas I, Parra-Llorca A, Sánchez-Illana A, Nuñez-Ramiro A, Kuligowski J, Cháfer-Pericás C, Cernada M, Escobar J, Vento M. Oxygen and oxidative stress in the perinatal period. *Redox Biol.* 2017;12:674–681.
71. Johnston MV. Neurbiology of hypoxic-ischemic injury in the developing brain. *Pediatr Res.* 2001;49:735–741.
72. Johnston MV. Selective vulnerability in the neonatal brain. *Ann Neurol.* 1998;44:155–156.
73. Chevalier M, Lory P, Mironneau C, Macrez N, Quignard JF. T-type CaV3.3 calcium channels produce spontaneous low-threshold action potentials and intracellular calcium oscillations. *Eur J Neurosci.* 2006;23:2321–2329.
74. McCord JM. Oxygen-derived free radicals in postischemic tissue injury. *N Engl J Med.* 1985;312:159–163.
75. Martin LJ. The apoptosis-necrosis cell death continuum in CNS development, injury and disease: contributions and mechanisms. In: Lo EH, Marwah J, editors. *Neuroprotection.* Boston: Prominent Press; 2002, pp. 379–412.

76. Fellman V, Raivo KO. Reperfusion injury as the mechanism of brain damage after perinatal asphyxia. *Pediatr Res.* 1997;41:599–606.
77. Jovandaric MZ, Nikolic TV, Milenkovic SJ, Otaševic BS, Bankovic VV, Ivanovski PI, Jesic MM. Lipids on the second day in ischemic and normoxemic term neonates. *Fetal Pediatr Pathol.* 2017;36:276–281.
78. Niki E, Yoshida Y, Saito Y, Noguchi N. Lipid peroxidation: mechanisms, inhibition, and biological effects. *Biochem Biophys Res Commun.* 2005;338:668–676.
79. Gaschler MM, Stockwell BR. Lipid peroxidation in cell death. *Biochem Biophys Res Commun.* 2017;482:419–425.
80. Rochette L, Dogon G, Rigal E, Zeller M, Cottin Y, Vergely C. Lipid peroxidation and iron metabolism: two corner stones in the homeostasis control of ferroptosis. *Int J Mol Sci.* 2022;24:449.
81. Arai H. Oxidative modification of lipoproteins. *Subcell Biochem.* 2014;77:103–114.
82. Matsuura E, Lopez LR. Are oxidazed LDL/β2-glycoprotein 1 complexes pathogenic antigens in autoimmune-mediated atherosclerosis? *Clin Dev Immunol.* 2004;11:103–111.
83. Leren TP. Sorting an LDL receptor with bound PCSK9 to intracellular degradation. *Atherosclerosis.* 2014;237:76–81.
84. Park YM, Febbraio M, Silverstein RL. CD36 modulates migration of mouse and human macrophages in response to oxidized LDL and may contribute to macrophage trapping in the arterial intima. *J Clin Invest.* 2009;119:136–145.
85. Vlassaks E, Nikiforou M, Strackx E, Hütten M, Bekers O, Gazzolo D, Li Volti G, Martinez-Martinez P, Kramer BW, Gavilanes AW. Acute and chronic immunomodulatory changes in rat liver after fetal and perinatal asphyxia. *J Dev Orig Health Dis.* 2014;5:98–108.
86. Gleissner CA, Leitinger N, Ley K. Effects of native and modified low-density lipoproteins on monocyte recruitment in atherosclerosis. *Hypertension.* 2007;50:276–283.
87. Lankin VZ, Tikhaze AK, Kapel'ko VI, Shepel'kova GS, Shumaev KB, Panasenko OM, Konovalova GG, Belenkov YN. Mechanisms of oxidative modification of low density lipoproteins under conditions of oxidative and carbonyl stress *Biochemistry.* 2007;72:1081–1090.

88. Chen B, Meng L, Shen T, Gong H, Qi R, Zhao Y, Sun J, Bao L, Zhao G. Thioredoxin attenuates oxidized low-density lipoprotein induced oxidative stress in human umbilical vein endothelial cells by reducing NADPH oxidase activity. *Biochem Biophys Res Commun.* 2017;490:1326–1333.
89. Mertens A, Holvoet P. Oxidized LDL and HDL: antagonists in atherothrombosis. *The FASEB J.* 2001;15:2073–2084.
90. León-Pedroza JI, González-Tapia LA, del Olmo-Gil E, Castellanos-Rodríguez D, Escobedo G, González-Chávez A. Low-grade systemic inflammation and the development of metabolic diseases: from the molecular evidence to the clinical practice. *Cir Cir.* 2015;83:543–551.
91. Bujo H, Saito Y. Modulation of smooth muscle cell migration by members of the low-densitylipoprotein receptor family. *Arterioscler Thromb Vasc Biol.* 2006;26:1246–1252.
92. Stoll G, Bendszus M. Inflammation and atherosclerosis novel insights into plaque formation and destabilization. *Stroke.* 2006;37:1923.
93. Kobayashi S, Faber N, Ohashi Y, Terashima M,Matsui K, Mori T, Fujita H, Awano K, et al. Interaction of oxidative stress and inflammatory response in coronary plaque instability: important role of C-reactive protein. *Arterioscler Thromb Vasc Biol.* 2003;23:1398.
94. Galle J, Hansen-Hagge T, Wanner C, Seibold S. Impact of oxidized low density lipoprotein onvascular cells. *Atherosclerosis.* 2006;185:219–226.
95. Diczfalusy U. Analysis of cholesterol oxidation products in biological samples. *J AOAC Int.* 2004;87:467–473.
96. Esterbauer H, Dieber-Rotheneder M, Waeg G, Striegl G, Jürgens G. Biochemical, structural, and functional properties of oxidized low-density lipoprotein. *Chem Res Toxicol.* 1990;3:77–92.
97. Ramy N, Al Sharany W, Mohamed MA, Madani H, Saleh E, Aly H. Lipid peroxides in the serum of asphyxiated neonates. *J Perinatol.* 2016;36:849–852.
98. Yoshida Y, Ito N, Shimakawa S, Niki E. Susceptibility of plasma lipids to peroxidation. *Biochem Biophys Res Commun.* 2003;305:747–753.
99. Wang J, Wu Z, Li D, Li N, Dindot SV, Sattereld MC, Bazer FW, Wu G. Nutrition, epigenetics, and metabolic syndrome. *Antioxid Redox Sign.* 2012;17:282–301.

100. Tarry-Adkins JL, Ozanne SE. Nutrition in early life and age-associated diseases. *Ageing Res Rev.* 2017;39:96–105.
101. Cropley JE, Eaton SA, Aiken A, Young PE, Giannoulatou E, Ho JWK, Buckland ME, Keam SP, Hutvagner G, Humphreys DT, Langley KG, Henstridge DC, Martin DIK, Febbraio MA, Suter CM. Male-lineage transmission of an acquired metabolic phenotype induced by grand-paternal obesity. *Mol Metab.* 2016;5:699–708.
102. Fernandez-Twinn DS, Hjort L, Novakovic B, Ozanne SE, Saffery R. Intrauterine programming of obesity and type 2 diabetes. *Diabetologia.* 2019;62:1789–1801.
103. Wang S, Qian J, Sun F, Li M, Ye J, Li M, Du M, Li D. Bidirectional regulation between 1st trimester HTR8/SVneo trophoblast cells and in vitro differentiated Th17/Treg cells suggest a fetal-maternal regulatory loop in human pregnancy. *Am J Reprod Immunol.* 2019;81:e13106.
104. Lentz LS, Stutz AJ, Meyer N, Schubert K, Karkossa I, von Bergen M, Zenclussen AC, Schumacher A. Human chorionic gonadotropin promotes murine Treg cells and restricts pregnancy-harmful proinflammatory Th17 responses. *Front Immunol.* 2022 Sep;20 13:989247.
105. Qian J, Zhang N, Lin J, Wang C, Pan X, Chen L, Li D, Wang L. Distinct pattern of Th17/Treg cells in pregnant women with a history of unexplained recurrent spontaneous abortion. *Biosci Trends.* 2018;12:157–167.
106. Jing M, Chen X, Qiu H, He W, Zhou Y, Li D, Wang D, Jiao Y, Liu A. Insights into the immunomodulatory regulation of matrix metalloproteinase at the maternal-fetal interface during early pregnancy and pregnancy-related diseases. *Front Immunol.* 2023;13:1067661.
107. Rassie K, Giri R, Joham AE, Teede H, Mousa A. Human placental lactogen in relation to maternal metabolic health and fetal outcomes: a systematic review and meta-analysis. *Int J Mol Sci.* 2022;23:15621.
108. Sibiak R, Gutaj P, Wender-Ozegowska E, Jankowski M, Mozdziak P, Kempisty B. Placental lactogen as a marker of maternal obesity, diabetes, and fetal growth abnormalities: current knowledge and clinical perspectives. *J Clin Med.* 2020;9:1142.

109. Costa M A. The endocrine function of human placenta: an overview. *Reprod Biomed Online.* 2016;32:14–43.
110. Napso T, Yong HEJ, Lopez-Tello J, Sferruzzi-Perri AN. The role of placental hormones in mediating maternal adaptations to support pregnancy and lactation. *Front Physiol.* 2018;9:1091.
111. Shende P, Desai D. Physiological and therapeutic roles of neuropeptide Y on biological functions. *Adv Exp Med Biol.* 2020;1237:37–47.
112. Yosten GLC, Haddock CJ, Harada CM, Almeida-Pereira G, Kolar GR, Stein LM, Hayes MR, Salvemini D, Samson WK. Past, present and future of cocaine- and amphetamine-regulated transcript peptide. *Physiol Behav.* 2021;235:113380.
113. Stern C, Schwarz S, Moser G, Cvitic S, Jantscher-Krenn E, Gauster M, Hiden U. Placental endocrine activity: adaptation and disruption of maternal glucose metabolism in pregnancy and the influence of fetal sex. *Int J Mol Sci.* 2021;22:12722.
114. Waage CW, Mdala I, Stigum H, Jenum AK, Birkeland KI, Shakeel N, Michelsen TM, Richardsen KR, Sletner L. Lipid and lipoprotein concentrations during pregnancy and associations with ethnicity. *BMC Pregnancy Childbirth.* 2022;22:246.
115. Gootjes DV, Posthumus AG, Wols DF, de Rijke YB, Roeters Van Lennep JE, Steegers EAP. Maternal lipid profile in pregnancy and embryonic size: a population-based prospective cohort study. *BMC Pregnancy Childbirth.* 2022;22:333.
116. Lippi G, Albiero A, Montagnana M, Salvagno GL, Scevarolli S, Franchi M, Guidi GC. Lipid and lipoprotein profile in physiological pregnancy. *Clin Lab.* 2007;53:173–177.
117. Roland MCP, Godang K, Aukrust P, Henriksen T, Lekva T. Low CETP activity and unique composition of large VLDL and small HDL in women giving birth to small-for-gestational age infants. *Sci Rep.* 2021;11:6213.
118. Loukidi-Bouchenak B, Lamri-Senhadji MY, Merzouk S, Merzouk H, Belarbi B, Prost J, Belleville J, Bouchenak M. Serum lecithin: cholesterol acyltransferase activity, HDL2 and HDL3 composition in hypertensive mothers and their small for gestational age newborns. *Eur J Pediatr.* 2008;167:525–532.
119. Chavan-Gautam P, Rani A, Freeman DJ. Distribution of fatty acids and lipids during pregnancy. *Adv Clin Chem.* 2018;84:209–239.

120. Perazzolo S, Hirschmugl B, Wadsack C, Desoye G, Lewis RM, Sengers BG. The influence of placental metabolism on fatty acid transfer to the fetus. *J Lipid Res.* 2017;58:443–454.
121. Barrett HL, Kubala MH, Scholz Romero K, Denny KJ, Woodruff TM, McIntyre HD, Callaway LK, Nitert MD. Placental lipases in pregnancies complicated by gestational diabetes mellitus (GDM). *PLoS One.* 2014;9:e104826.
122. Grace MR, Vladutiu CJ, Nethery RC, Siega-Riz AM, Manuck TA, Herring AH, Savitz D, Thorp JT. Lipoprotein particle concentration measured by nuclear magnetic resonance spectroscopy is associated with gestational age at delivery: a prospective cohort study. *BJOG.* 2018;125:895–903.
123. Chassen S, Jansson T. Complex, coordinated and highly regulated changes in placental signaling and nutrient transport capacity in IUGR. *Biochim Biophys Acta Mol Basis Dis.* 2020;1866:165373.
124. Kopčeková J, Kolesárová A, Schwarzová M, Kováčik A, Mrázová J, Gažarová M, Lenártová P, Chlebo P, Kolesárová A. Phytonutrients of bitter apricot seeds modulate human lipid profile and LDL subfractions in adults with elevated cholesterol levels. *Int J Environ Res Public Health.* 2022;19:857.
125. Ivanova EA, Myasoedova VA, Melnichenko AA, Grechko AV, Orekhov AN. Small dense low-density lipoprotein as biomarker for atherosclerotic diseases. *Oxid Med Cell Longev.* 2017;2017:1273042.
126. Berberich AJ, Hegele RA. A modern approach to dyslipidemia. *Endocr Rev.* 2022;43:611–653.
127. Feitosa ACR, Barreto LT, Silva IMD, Silva FFD, Feitosa GS Filho. Impact of the use of different diagnostic criteria in the prevalence of dyslipidemia in pregnant women. *Arq Bras Cardiol.* 2017;109:30–38.
128. Tani S. The ratio of triglyceride to high-density lipoprotein cholesterol as an indicator of risk stratification for atherosclerotic cardiovascular disease in a clinical setting. *Intern Med.* 2020;59:2639–2640.
129. Moriyama K. The association between the triglyceride to high-density lipoprotein cholesterol ratio and low-density lipoprotein subclasses. *Intern Med.* 2020;59:2661–2669.

130. Cibickova L, Schovanek J, Karasek D. Changes in serum lipid levels during pregnancy in women with gestational diabetes: a narrative review. *Biomed Pap Med Fac Palacky Univ Olomouc.* 2021;165:8–12.
131. Huang JK, Lee HC. Emerging evidence of pathological roles of very-low-density lipoprotein (VLDL). *Int J Mol Sci.* 2022;23:4300.
132. Wang Z, Peng Y, Mao S, Zhang L, Guo Y. The correlation between blood-lipid ratio in the first trimester and large-for-gestational-age infants. *Lipids Health Dis.* 2023;22:18.
133. Mauri M, Calmarza P, Ibarretxe D. Dyslipemias and pregnancy, an update. *Clin Investig Arterioscler.* 2021;33:41–52.
134. Barbour LA, Hernandez TL. Maternal lipids and fetal overgrowth: making fat from fat. *Clin Ther.* 2018;40:1638–1647.
135. von Eckardstein A, Nordestgaard BG, Remaley AT, Catapano AL. High-density lipoprotein revisited: biological functions and clinical relevance. *Eur Heart J.* 2023;44:1394–1407.
136. Rohatgi A, Westerterp M, von Eckardstein A, Remaley A, Rye KA. HDL in the 21st century: a multifunctional roadmap for future HDL research. *Circulation.* 2021 8;143:2293–2309.
137. Aleksenko L, Quaye IK. Pregnancy-induced cardiovascular pathologies: importance of structural components and lipids. *Am J Med Sci.* 2020;360:447–466.
138. Wang CD, Ma Y, Li SW, Li QX, Zhang L, Lin LL, Huang BB, Jiang YS. Analysis of risk factors for gestational diabetes mellitus in elderly multipara women in the next pregnancy. *Zhonghua Yi Xue Za Zhi.* 2022;102:418–422.
139. Kapoor B, Kapoor D, Gautam S, Singh R, Bhardwaj S. Dietary polyunsaturated fatty acids (PUFAs): uses and potential health benefits. *Curr Nutr Rep.* 2021;10:232–242.
140. Jovandaric MZ, Milenkovic SJ, Babovic IR, Babic S, Dotlic J. The effect of glucose metabolism and breastfeeding on the intestinal microbiota of newborns of women with gestational diabetes mellitus. *Medicina (Kaunas).* 2022;58:413.
141. Hardy MLM, Day ML, Morris MB. Redox regulation and oxidative stress in mammalian oocytes and embryos developed in vivo and in vitro. *Int J Environ Res Public Health.* 2021;18:11374.

142. Hussain T, Murtaza G, Metwally E, Kalhoro DH, Kalhoro MS, Rahu BA, Sahito RGA, Yin Y, Yang H, Chughtai MI, Tan B. The role of oxidative stress and antioxidant balance in pregnancy. *Mediators Inflamm.* 2021;2021:9962860.
143. Joo EH, Kim YR, Kim N, Jung JE, Han SH, Cho HY. Effect of endogenic and exogenic oxidative stress triggers on adverse pregnancy outcomes: preeclampsia, fetal growth restriction, gestational diabetes mellitus and preterm birth. *Int J Mol Sci.* 2021;22:10122.
144. Pospíšil P, Prasad A, Rác M. Mechanism of the formation of electronically excited species by oxidative metabolic processes: role of reactive oxygen species. *Biomolecules.* 2019;9:258.
145. Kumar R, Jafri MS. Computational modeling of mitochondria to understand the dynamics of oxidative stress. *Methods Mol Biol.* 2022;2497:363–422.
146. Yin Y, Shen H. Common methods in mitochondrial research (review). *Int J Mol Med.* 2022;50:126.
147. Rottenberg H, Hoek JB. The path from mitochondrial ROS to aging runs through the mitochondrial permeability transition pore. *Aging Cell.* 2017;16:943–955.
148. Villalpando-Rodriguez GE, Gibson SB. Reactive oxygen species (ROS) regulates different types of cell death by acting as a rheostat. *Oxid Med Cell Longev.* 2021;2021:9912436.
149. Zhao RZ, Jiang S, Zhang L, Yu ZB. Mitochondrial electron transport chain, ROS generation and uncoupling (review). *Int J Mol Med.* 2019;44:3–15.
150. Sousa JS, D'Imprima E, Vonck J. Mitochondrial respiratory chain complexes. *Subcell Biochem.* 2018;87:167–227.
151. Sies H. Hydrogen peroxide as a central redox signaling molecule in physiological oxidative stress: oxidative eustress. *Redox Biol.* 2017;11:613–619.
152. Niki E. Oxidative stress and antioxidants: distress or eustress? *Arch Biochem Biophys.* 2016;595:19–24.
153. Jovandaric MZ, Dokic M, Babovic IR, Milicevic S, Dotlic J, Milosevic B, Culjic M, Andric L, Dimic N, Mitrovic O, Beleslin A, Nikolic J, Jestrovic Z, Babic S. The significance of COVID-19 diseases in lipid metabolism pregnancy women and newborns. *Int J Mol Sci.* 2022;23:15098.
154. Rubarth BL, Quinn J. Respiratory development and respiratory distress syndrome. *Neonatal Netw.* 2015;34:231–238.

155. Jovandaric MZ, Milenkovic SJ, Dotlic J, Babovic IR, Jestrovic Z, Milosevic B, Culjic M, Babic S. Neonatal pneumothorax outcome in preterm and term newborns. *Medicina (Kaunas)*. 2022;58:965.
156. Conrad M, Kagan VE, Bayir H, Pagnussat GC, Head B, Traber MG, Stockwell BR. Regulation of lipid peroxidation and ferroptosis in diverse species. *Genes Dev*. 2018;32:602–619.
157. Nandi A, Yan LJ, Jana CK, Das N. Role of catalase in oxidative stress- and age-associated degenerative diseases. *Oxid Med Cell Longev*. 2019;2019:9613090.
158. Thilaganathan B. Placental syndromes: getting to the heart of the matter. *Ultrasound Obstet Gynecol*. 2017;49:7–9.
159. Beharier O, Tyurin VA, Goff JP, Guerrero-Santoro J, Kajiwara K, Chu T, Tyurina YY, St Croix CM, Wallace CT, Parry S, Parks WT, Kagan VE, Sadovsky Y. PLA2G6 guards placental trophoblasts against ferroptotic injury. *Proc Natl Acad Sci U S A*. 2020;117:27319–27328.
160. Iuchi K, Takai T, Hisatomi H. Cell death via lipid peroxidation and protein aggregation diseases. *Biology (Basel)*. 2021;10:399.
161. Collin F. Chemical basis of reactive oxygen species reactivity and involvement in neurodegenerative diseases. *Int J Mol Sci*. 2019;20:2407.
162. Ayala A, Munoz MF, Arguelles S. Lipid peroxidation: production, metabolism, and signaling mechanisms of malondialdehyde and 4-hydroxy-2-nonenal. *Oxid Med Cell Longev*. 2014;2014:360438.
163. Iuchi K. Manipulation of cell fate by fatty acids and oxidized fatty acids. *Agric Biotechnol*. 2021;5:38–42.
164. Kowalczyk P, Sulejczak D, Kleczkowska P, Bukowska-Ośko I, Kucia M, Popiel M, Wietrak E, Kramkowski K, Wrzosek K, Kaczyńska K. Mitochondrial oxidative stress: a causative factor and therapeutic target in many diseases. *Int J Mol Sci*. 2021;22:13384.
165. Girotti AW, Korytowski W. Cholesterol as a natural probe for free radical–mediated lipid peroxidation in biological membranes and lipoproteins. *J Chromatogr B Analyt Technol Biomed Life Sci*. 2016;1019:202–209.
166. Itabe H. Oxidized low-density lipoproteins: what is understood and what remains to be clarified. *Biol Pharm Bull*. 2003;26:1–9.

167. Wu J, He Z, Gao X, Wu F, Ding R, Ren Y, Jiang Q, Fan M, Liang C, Wu Z. Oxidized high-density lipoprotein impairs endothelial progenitor cells' function by activation of CD36-MAPK-TSP-1 pathways. *Antioxid Redox Signal.* 2015;22:308–324.
168. Biermann M, Maueröder C, Brauner JM, Chaurio R, Janko C, Herrmann M, Muñoz LE. Surface code: biophysical signals for apoptotic cell clearance. *Phys Biol.* 2013;10:065007.
169. Hinkovska-Galcheva V, VanWay SM, Shanley TP, Kunkel RG. The role of sphingosine-1-phosphate and ceramide-1-phosphate in calcium homeostasis. *Curr Opin Investig Drugs.* 2008;9:1192–1205.
170. Bayés García R, Quiles Guardia P, Neira Antonio B, Molina Font J. Effect of perinatal hypoxia on blood triglycerides and total cholesterol including high density lipoproteins. *An Esp Pediatr.* 1988;29:15–22.
171. Sharpe LJ, Cook EC, Zelcer N, Brown AJ. The UPS and downs of cholesterol homeostasis. *Trends Biochem Sci.* 2014;39:527–535.

Chapter 7

Lipids: A Predictive Parameter of Therapeutic Hypothermia (Cooling Therapy) in Asphyxiated Newborns

Miljana Z. Jovandaric

7.1 Introduction

Hypoxia-ischemia is a rare phenomenon that can occurs before or during childbirth after a period of reduced placental and/or fetal blood flow. After recovery, there may be a phase of reperfusion and secondary loss of energy 6–48 hours after the initial insult. Therapeutic hypothermia can be offered to infants at risk of developing encephalopathy if identified within 6 hours of the therapeutic period and should be instituted as soon as possible for eligible infants. In addition, the physician must pay close attention to supportive measures

DOI: 10.1201/9781003346692-7

such as avoidance of hyperthermia, as well as comprehensive treatment of clinical or electrographic seizures, blood pressure, blood glucose, and carbon dioxide levels (1).

Therapeutic hypothermia is a form of neuroprotective therapy in asphyxiated newborns that has been in use for the last 20 years. Before hypothermia, the therapy of hypoxic-ischemic encephalopathy (HIE) as a direct consequence of perinatal asphyxia was only supportive, and severe consequences could not be avoided, namely immediate death and delayed severe neurological sequelae in surviving newborns. The neuroprotective value of hypothermia has been known for a long time and is applied to cardiac and neurosurgical patients. In the last 10 years, intensive work has been done to create a protocol for the application of therapeutic hypothermia in asphyxiated newborns, which resulted in the introduction of therapeutic hypothermia in the guidelines for cardiopulmonary resuscitation of asphyxiated newborns from 2010 (2).

Perinatal asphyxia is the term used for a disorder of respiratory gas exchange in the placental circulation that results in low perfusion of fetal tissues, hypoxia, hypercapnia, and acidosis. The consequences on fetal cerebral and systemic blood flow can be significant and lead to multiple-organ failure (3–5). One of the most severe consequences of significant perinatal asphyxia is hypoxic-ischemic brain injury, the HIE. The frequency of hypoxic-ischemic encephalopathy in developed countries is 1 to 2 cases per 1,000 full-term live births (6). Part of the cause is intrapartum complications: abruption of the placenta, rupture of the uterus, embolization with amniotic fluid, prolapse, and avulsion of the umbilical cord. Great importance is attributed to antenatal risk factors such as severe preeclampsia, maternal hypertension and thyroid disease, and vaginal bleeding during pregnancy (7). Most encephalopathies have no obvious cause. HIE represents an acute injury and death of neurons and can result in immediate death as well as delayed numerous neurological sequelae that are diagnosed in infancy and later life, such as cerebral palsy and psychomotor and cognitive retardation. The risk of death in newborns with moderate encephalopathy is 10%, and the frequency of neurological damage in survivors is 30%. In newborns with severe encephalopathy, 60% are fatal, and almost 100% of survivors have some form of disability (8).

Until a few years ago, the possible therapy for HIE was only supportive, in intensive care units for newborns. In the last 20 years, a new possibility of treatment in the form of hypothermia has gradually begun to be introduced. Namely, experiments on animals showed that moderate hypothermia has a partial neuroprotective effect after brain ischemia (9). Postischemic hypothermia was associated with less severe stages of encephalopathy compared to normothermia. It has been proven that neuronal injury in HIE is a continuous process and not an isolated one-time event (10). Primary cellular injury is a direct consequence of hypoxia and ischemia. Low cerebral concentrations of energy-rich compounds adenosine triphosphate and creatine phosphorus were observed during the primary injury. This drop in energy causes the creation of cytotoxic edema and neuronal injury. The next phase is the reperfusion phase, in which the cellular edema recedes in about 30 minutes. Then follows a latent phase in which the oxidative cerebral energy metabolism is very close to normal. Experiments on animal models suggest that the latent phase lasts several hours. After the latent phase, 6–15 hours after acute asphyxia, secondary injury develops with secondary cytotoxic edema, extracellular accumulation of excitatory amino acids, generation of free radicals, induction of apoptosis and inflammatory activity, and final breakdown of oxidative metabolism and neuronal death. Clinically, the secondary phase is presented by convulsions (3, 11). The idea of hypothermia therapy in asphyxiated newborns is to cool down the affected newborn in the latent phase of HIE development and thus prevent the secondary phase and secondary neuron injury (12).

7.1.1 Therapeutic Hypothermia (Cooling Therapy) for Newborns

Hypothermia reduces brain injury by affecting several biological processes. It reduces vasogenic edema, the bleeding that often accompanies asphyxia and neutrophil infiltration. It limits the release of excitatory amino acids, the accumulation of intracellular calcium, and the creation of free oxygen radicals. It protects cells and organelles from oxidative damage during the reperfusion phase. By reducing the consumption of glucose and oxygen, it also helps maintain cerebral metabolism. It also acts on anti-inflammatory cytokines, the coagulation cascade, and caspase enzymes that participate in the programmed death of neurons (11, 13). There are numerous questions that arise when considering hypothermia therapy, starting with the issue of safety. The first use of hypothermia as a therapy after perinatal asphyxia was applied in 1959 (14). The fact is that experiments on animals and later clinical research on therapeutic hypothermia can be found in large numbers in the literature for the past 20 years. The reason for the time discrepancy between the knowledge of the potential benefit of hypothermia for a given indication and the more intensive evidence-based research can be found in the knowledge of the thermoregulation of newborns. It has been known for a long time that the control of thermal conditions is of vital importance for the newborn due to their weak ability for internal thermoregulation. Hypothermia in newborns can lead to severe consequences and even fatal outcomes. The question is the boundary between beneficial hypothermia and hypothermia that can end fatally (15). It has been established that therapeutic hypothermia is effective in reducing the risk of death in infants with moderate to severe hypoxic-ischemic encephalopathy. In addition, both selective head cooling and whole body cooling methods are effective in reducing mortality in infants with this condition. The pooled relative risk of death among asphyxiated infants who received cooling therapy was found to be nearly 26% lower compared to those who did not. They did not receive cooling therapy (16).

In evaluating the effectiveness of hypothermia as a neuroprotective strategy, there are three critical variables: the time of initiation of hypothermia after birth, the duration of controlled hypothermia, and the level of hypothermia (target body temperature of the newborn undergoing therapy) (17). In addition, there remains the challenge of selecting newborns who will benefit from hypothermia without exposing newborns who are not good candidates and whom hypothermia will not benefit but will harm (18).

The first step in the implementation of therapeutic hypothermia is good patient selection. Given that the therapeutic window is very narrow (subcooling should be started within 6 hours of birth), the selection must be made already at birth, which is not an easy task. The main inclusion criteria are moderate or severe encephalopathy and indicators of peripartum hypoxia-ischemia. The conditions that must be met in order for a child to be considered for hypothermia therapy are the gestational age of the newborn of 36 weeks or more and the possibility of starting the therapy within 6 hours after birth. All studies show that included patients must be born at term or close to term. Most of the large studies included children with a gestational age of 36 or more weeks (19), while in two studies they set the minimum gestational age to be 37 weeks (20). Jacobs et al. included in their research children from 35 weeks of gestation (12). In late preterm infants with confirmed asphyxia, magnetic resonance imaging results are consistent with a two-phase model of neuronal death. These studies show normal cerebral oxidative metabolism shortly after birth followed by “secondary energy failure,” the degree of which predicts outcome (mortality and neurodevelopmental outcome at 1 and 4 years of age) (21).

Therefore, a therapeutic “window of opportunity” exists in the interval after resuscitation of an asphyxiated newborn before the secondary phase of disturbed energy metabolism and injury (22). The first step in the implementation of therapeutic hypothermia is good patient selection. Given that the therapeutic window is very narrow (subcooling should be started within 6 hours of birth), the selection must be made already at birth, which is not an easy task. The main inclusion criteria are moderate or severe encephalopathy and indicators of peripartum hypoxia-ischemia. The conditions that must be met in order for a child to be considered for hypothermia therapy are the gestational age of the newborn of 36 weeks or more and the possibility of starting the therapy within 6 hours after birth. All studies show that included patients must be born at term or close to term. Most of the large studies included children with a gestational age of 36 or more weeks (23), while in two studies they set the minimum gestational age to be 37 weeks (20). Jacobs et al. included in their research children with a gestational age of 35 weeks (24).

Induction of hypothermia by the whole body cooling (WBC) method must start within 6 hours of birth. As the name suggests, this method cools the whole body of the newborn. The target temperature of systemic hypothermia is between 33°C and 34°C (23). In one study from 2005, the temperature achieved was on average 0.5°C lower (target temperature 32.5–33.5°C) (25). Cooling by the WBC method lasts 72 hours. Cooling using the WBC method for 48 hours is also shown in the literature (19). During therapeutic hypothermia, they monitor all vital functions (blood pressure, pulse, respiration rate, oxygen saturation) and routinely monitor other parameters such as arterial blood gas exchange, coagulation, complete blood count, electrolytes, urea and creatinine, and urine volume. After 72 hours, rewarming is carried out in such a way that the temperature of the newborn rises by 0.5°C in an hour, no more. In one study, the warming time was extended in such a way that the temperature increased by 0.5°C in 2 hours (24). Cerebrospinal fluid biomarkers most commonly used to predict the most severe forms of HIE include creatine kinase, xanthine oxidase, vascular endothelial growth factor, neuron-specific enolase, superoxide dismutase, and malondialdehyde (26).

7.2 Lipids as a Parameter for Cooling Therapy Using the Whole-Body Cooling Method

Loss of cell membrane function is the primary event in the occurrence of cell damage in hypoxic conditions. Asphyxia leads to the oxidation of lipoproteins, where there is a correlation between the concentration of lipid hydroperoxides in the plasma and a more severe degree of hypoxia, that is, the manifestation of the consequences of asphyxia. The most frequently oxidized lipids are in low-density lipoproteins (LDL), and LDL is the main source of lipids that accumulate during ischemia. Excessive lipid oxidation alters the physical properties of cell membranes and can cause covalent modification of proteins and nucleic acids (27). As highly reactive compounds, lipid peroxides are also able to propagate further ROS generation or to degrade into reactive compounds capable of cross-linking DNA and proteins. Lipid peroxides can be divided into two general classes: lipid endoperoxides and lipid hydroperoxides. Lipid endoperoxides are a key intermediate in the formation of prostaglandins, and their importance in inflammation and HIE is very significant (28). The relative sensitivity of lipids to oxidation depends on their structure. Lipid hydroperoxides are formed as primary products, but they are substrates for different enzymes and thus undergo different secondary processes (29). Lipid peroxidation, especially in nerve cells, is catalyzed by iron released from hemoglobin, transferrin, and ferritin, the release of which is induced by tissue acidosis and free oxygen radicals (30).

Ferroptosis is a non-apoptotic form of iron-dependent regulated cell death characterized by the accumulation of lipid peroxides (31, 32). Ferroptosis was discovered through a high-throughput screening of lethal molecules, which led to the initial discovery of two small molecules that induce ferroptosis: erasing and RSL3. Both of these molecules exert their lethal effects by compromising the antioxidant capacity of cells. Erastin inhibits the Ksc^- system, a transmembrane cystine/glutamate antiporter that supplies cysteine for the synthesis of the antioxidant tripeptide glutathione. RSL3 covalently inhibits glutathione peroxidase 4 (GPx4), which reduces lipid peroxides to the corresponding hydroxide, using glutathione as a cofactor. Monitoring oxidative stress in ferroptotic cells showed an increase in lipid peroxidation without an increase in other ROS such as superoxide (33).

Oxidatively modified LDL particles (Oh-LDL) can also be created by direct oxidation, which mainly takes place on unsaturated fatty acids under the influence of superoxide and peroxynitrite radicals, or by the transfer of already oxidized cellular lipids to LDL. In doing so, lipid peroxides are formed, which enter into a new chain reaction, as well as conjugated dienes and aldehyde products. These changes are preceded by a "hesitation period" of aldehyde oxidation, i.e., a latent period in which the antioxidants in LDL are consumed, after which the oxidation of fatty acids can begin (34, 35).

At the same time, degraded LDL becomes a ligand for macrophage receptors, and in this way monocytes are attracted to the site of the lesion, "take" lipoproteins, and gradually turn into foam cells. When LDL deposits exceed a critical threshold, excessive activation of complement and macrophages creates the conditions for chronic inflammation (36).

Lipid peroxidation (LPO) as the most pronounced negative phenomenon in the action of free radicals represents an autocatalytic, progressive, and most often irreversible process. The initiation of lipid peroxidation process is initiated by any oxidant capable of borrowing a hydrogen atom from the double bond of an unsaturated fatty acid. The most responsible system for the oxidative modification of the LDL particle is the cellular lipoxygenase system (37). Oxidative modification of LDL is possible through peroxynitrite, which is formed in the reaction of nitrogen monoxide (NO) and superoxide, which are produced to an increased extent in conditions of endothelial activation, given that the induction of NO synthase and endothelial NADPH oxidase (38). Oxidized LDL (Oh-LDL) represents a modified form of LDL, whereby the oxidative modification takes place via SR. Namely, in the course of normal metabolic activity, there is a controlled creation of free radicals (atoms or molecules that have one unpaired electron, which is why they are very unstable and reactive). When metabolic disorders occur, free radicals are created uncontrollably and lead to significant damage to macromolecules. Free radicals react with lipids, proteins, and nucleic acids, leading to their oxidative modification. When lipids are damaged by free radicals, lipid free radicals and lipid peroxides are formed. It is considered that the main agent of LDL oxidation is superoxide anion, under the influence of which both the lipid and protein components of LDL particles are modified (39).

Oh-LDL is characterized by reactive aldehydes, reduced cholesterol content with the reduced number of reactive amino groups on apo B-100, elevated oxysterols, and fragmentation of apo B-100. Peroxidation of polyunsaturated fatty acids leads to the formation of highly reactive aldehydes that have a cytotoxic effect. Oxidation of apo B-100 leads to the loss of antigenic tolerance with the appearance of antibodies to modified LDL in the intima of arteries. As a result, there is a reduced recognition of the LDL receptor by the LDL particle and its increased binding to free radicals. There are no autoregulatory mechanisms for this pathway, so the cells that use it are not protected from uncontrolled cholesterol accumulation. Such cells include smooth muscle cells and monocytic macrophage cells (40).

There are two stages in the oxidative modification of the LDL particle. Minimally oxidized LDL occurs when only the lipid component is slightly oxidized, in contrast to complete oxidation of LDL when there are changes in the structure of apo B-100. Modification of lysine residues on apo B-100 changes its structure and leads to the removal of LDL particles via receptors on macrophages (41).

Ox-LDL stimulates endothelial cells to produce monocyte chemotactic factor, which leads to the accumulation of monocytes in the intima of blood vessels and their differentiation into macrophages. Macrophages can secrete Ox-LDL and superoxide anion, thereby becoming cytotoxic to the environment and further damaging the endothelium (42). Considering that triglycerides are esters of cholesterol and free fatty acids, under conditions of hypoxia, oxidation of lipids, i.e., polyunsaturated fatty acids (PUFA) and cholesterol, occurs (43). The presence of cholesterol in the brain is essential for the maturation of neurons. The most frequently oxidized lipids are low-density lipoproteins (LDL). This susceptibility to oxidation can be attributed to the physical properties of the membrane and the outer surface of the nucleus, where the two lipid envelopes are located. The cell membrane is most susceptible to peroxidation as a consequence of the high content of unsaturated fatty acids (44). Lipid peroxidation captures lipid cholesterol and apolipoprotein HDL particles, reducing the concentration of I HDL cholesterol, although it has been proven in animal experiments that oxidized HDL particles have a protective effect on the vascular wall and cells that are susceptible to lipid peroxidation processes (45).

7.3 Conclusion

Since lipid peroxidation occurs during asphyxia, the severity of asphyxia can be determined by determining the concentration of lipids, total cholesterol, LDL, HDL, and triglycerides, and lipids are the modality for sending a newborn to therapeutic hypothermia (cooling therapy).

References

1. Kasdorf E, Perlman JM. Strategies to prevent reperfusion injury to the brain following intrapartum hypoxia-ischemia. *Semin Fetal Neonatal Med.* 2013;18:379–384.
2. Abate BB, Bimerew M, Gebremichael B, Mengesha Kassie A, Kassaw M, Gebremeskel T, Bayih WA. Effects of therapeutic hypothermia on death among asphyxiated neonates with hypoxic-ischemic encephalopathy: a systematic review and meta-analysis of randomized control trials. *PLoS One.* 2021;16:e0247229.
3. Cornette L. Therapeutic hypothermia in neonatal asphyxia. *Faxts Views Vis Obgyn.* 2012;4:133–139.
4. Gancia P, Pomero G. Therapeutic hypothermia in the prevention of hypoxic-ischaemic encephalopathy: new categories to be enrolled. *J Matern Fetal Neonatal Med.* 2012;25 Suppl 4:94–96.
5. Galvao TF, Silva MT, Marques MC, de Oliveira ND, Pereira MG. Hypothermia for perinatal brain hypoxia-ischemia in different resource settings: a systematic review. *J. Trop. Pediatr.* 2013;59:453–459.
6. Kapadia VS, Chalak LF, DuPont TL, Rollins NK, Brion LP, Wyckoff MH. Perinatal asphyxia with hyperoxemia within the first hour of life is associated with moderate to severe hypoxic-ischemic encephalopathy. *J Pediatr.* 2013;163:949–954.

7. Magalhães M, Rodrigues FP, Chopard MR, Melo VC, Melhado A, Oliveira I, Gallacci CB, Pachi PR, Lima Neto TB. Neuroprotective body hypothermia among newborns with hypoxic ischemic encephalopathy: three-year experience in a tertiary university hospital. A retrospective observational study. *Sao Paulo Med J.* 2015;133:314–319.
8. Shankaran S, Laptook AR, Pappas A, McDonald SA, Das A, Tyson JE, Poindexter BB, Schibler K, Bell EF, Heyne RJ, Pedroza C, Bara R, Van Meurs KP, Huitema CMP, Grisby C, Devaskar U, Ehrenkranz RA, Harmon HM, Chalak LF, DeMauro SB, Garg M, Hartley-McAndrew ME, Khan AM, Walsh MC, Ambalavanan N, Brumbaugh JE, Watterberg KL, Shepherd EG, Hamrick SEG, Barks J, Cotten CM, Kilbride HW, Higgins RD, Eunice Kennedy Shriver National Institute of Child Health and Human Development Neonatal Research Network. Effect of depth and duration of cooling on death or disability at age 18 months among neonates with hypoxic-ischemic encephalopathy: a randomized clinical trial. *JAMA.* 2017;318:57–67.
9. Lin Z, Yu H, Lin J, Chen S, Liang Z, Zhang Z. Mild hypothermia via selective head cooling as neuroprotective therapy in term neonates with perinatal asphyxia: an experience from a single neonatal intensive care unit. *J. Perinatol.* 2006;26:180–184.
10. AlKharfy TM. Induced hypothermia to treat neonatal hypoxic-ischemic encephalopathy. *Review of literature with meta-analysis and development of national protocol Neurosciences (Riyadh).* 2013;18:18–26.
11. Tagin MA, Woolcott CG, Vincer MJ, Whyte RK, Stinson DA. Hypothermia for neonatal hypoxic ischemic encephalopathy: an updated systematic review and meta-analysis. *Arch Pediatr Adolesc Med.* 2012;166:558–566.
12. Jacobs SE, Berg M, Hunt R, Tarnow-Mordi WO, Inder TE, Davis PG. Cooling for newborns with hypoxic ischaemic encephalopathy. *Cochrane Database Syst Rev.* 2013;2013(1):CD003311.
13. Zhu JJ, Yu BY, Huang XK, He MZ, Chen BW, Chen TT, Fang HY, Chen SQ, Fu XQ, Li PJ, Lin ZL, Zhu JH. Neferine protects against hypoxic-ischemic brain damage in neonatal rats by suppressing NLRP3-mediated inflammasome activation. *Oxid Med Cell Longev.* 2021;2021:6654954.

14. Gunn AJ, Gluckman PD, Gunn TR. Selective head cooling in newborn infants after perinatal asphyxia: a safety study. *Pediatrics*. 1998;102:885–892.
15. Korf JM, McCullough LD, Caretti V. A narrative review on treatment strategies for neonatal hypoxic ischemic encephalopathy. *Transl Pediatr*. 2023;12:1552–1571.
16. Edwards AD, Brocklehurst P, Gunn AJ, Halliday H, Juszczak E, Levene M, Strohm B, Thoresen M, Whitelaw A, Azzopardi D. Neurological outcomes at 18 months of age after moderate hypothermia for perinatal hypoxic ischaemic encephalopathy: synthesis and meta-analysis of trial data. *BMJ*. 2010;340:c363.
17. Laptook AR, Corbett RJT, Sterett R, Burns DK, Garcia D, Tollefsbol G. Modest hypothermia provides partial neuroprotection when used for immediate resuscitation after brain ischemia. *Pediatr Res*. 1997;42:17–23.
18. Natarajan G, Pappas A, Shankaran S. Outcomes in childhood following therapeutic hypothermia for neonatal hypoxic-ischemic encephalopathy (HIE). *Semin Perinatol*. 2016;40:549–555.
19. Azzopardi D, Robertson NJ, Cowan FM, Rutherford MA, Rampling M, Edwards AD. Pilot study of treatment with whole body hypothermia for neonatal encephalopathy. *Pediatrics*. 2000;106:684–694.
20. Zhou WH, Cheng GQ, Shao XM, Liu XZ, Shan RB, Zhuang DY, Zhou CL, Du LZ, Cao Y, Yang Q, Wang LS; China Study Group. Selective head cooling with mild systemic hypothermia after neonatal hypoxic-ischemic encephalopathy: a multicenter randomized controlled trial in China. *J Pediatr*. 2010;157(3):367–372, 372.e1–3.
21. Roth SC, Baudin J, Cady E, Johal K, Townsend JP, Wyatt JS, Reynolds EO, Stewart AL. Relation of deranged neonatal cerebral oxidative metabolism with neurodevelopmental outcome and head circumference at 4 years. *Dev Med Child Neurol*. 1997;39:718–725.
22. Denihan NM, Boylan GB, Murray DM. Metabolomic profiling in perinatal asphyxia: a promising new field. *Biomed Res Int*. 2015;2015:254076.
23. Azzopardi DV, Strohm B, Edwards AD, Dyet L, Halliday HL, Juszczak E, Kapellou O, Levene M, Marlow N, Porter E,

Thoresen M, Whitelaw A, Brocklehurst P; TOBY Study Group. Moderate hypothermia to treat perinatal asphyxial encephalopathy. *N Engl J Med.* 2009;361:1349–1358.

24. Jacobs SE, Morley CJ, Inder TE, Stewart MJ, Smith KR, McNamara PJ, Wright IM, Kirpalani HM, Darlow BA, Doyle LW, Infant Cooling Evaluation Collaboration. Whole-body hypothermia for term and near-term newborns with hypoxic-ischemic encephalopathy: a randomized controlled trial. *Arch Pediatr Adolesc Med.* 2011;165:692–700.
25. Eicher DJ, Wagner CL, Katikaneni LP, Hulsey TC, Bass WT, Kaufman DA, Horgan MJ, Languani S, Bhatia JJ, Givelichian LM, Sankaran K, Yager JY. Moderate hypothermia in neonatal encephalopathy: safety outcomes. *Pediatr Neurol.* 2005;32:18–24.
26. Shi Z, Luo K, Deol S, Tan S. A systematic review of noninflammatory cerebrospinal fluid biomarkers for clinical outcome in neonates with perinatal hypoxic brain injury that could be biologically significant. *J Neurosci Res.* 2022;100:2154–2173.
27. Gaschler MM, Stockwell BR. Lipid peroxidation in cell death. *Biochem Biophys Res Commun.* 2017;482:419–425.
28. Higdon A, Diers AR, Oh JY, Landar A, Darley-Usmar VM. Cell signalling by reactive lipid species: new concepts and molecular mechanisms. *Biochem J.* 2012;442:453–464.
29. Marisa Repetto JS, Alberto Boveris. Lipid peroxidation: chemical mechanism, biological implications and analytical determination. In: Catala DA, editor. *Lipid Peroxiation.* London, UK, InTech; 2012.
30. Yu T, Kui LQ, Ming QZ. Effect of asphyxia on non-protein-bound iron and lipid peroxidation in newborn infants. *Dev Med Child Neurol.* 2003;45:24–27.
31. Friedmann Angeli JP, Schneider M, Proneth B, Tyurina YY, Tyurin VA, Hammond VJ, Herbach N, Aichler M, Walch A, Eggenhofer E, Basavarajappa D, Radmark O, Kobayashi S, Seibt T, Beck H, Neff F, Esposito I, Wanke R, Forster H, Yefremova O, Heinrichmeyer M, Bornkamm GW, Geissler EK, Thomas SB, Stockwell BR, O'Donnell VB, Kagan VE, Schick JA, Conrad M. Inactivation of the ferroptosis regulator Gpx4 triggers acute renal failure in mice. *Nat Cell Biol.* 2014;16:1180–1191.

32. Dixon SJ, Lemberg KM, Lamprecht MR, Skouta R, Zaitsev EM, Gleason CE, Patel DN, Bauer AJ, Cantley AM, Yang WS, Morrison B, 3rd, Stockwell BR. Ferroptosis: an iron-dependent form of nonapoptotic cell death. *Cell.* 2012;149:1060–1072.
33. Yang WS, Kim KJ, Gaschler MM, Patel M, Shchepinov MS, Stockwell BR. Peroxidation of polyunsaturated fatty acids by lipoxygenases drives ferroptosis. *Proc Natl Acad Sci U S A.* 2016;113:E4966–4975.
34. Su LJ, Zhang JH, Gomez H, Murugan R, Hong X, Xu D, Jiang F, Peng Z-Y. Reactive oxygen species-induced lipid peroxidation in apoptosis, autophagy, and ferroptosis. *Oxid Med Cell Longev.* 2019;2019:5080843.
35. Kianmehr A, Qujeq D, Bagheri A, Mahrooz A. Oxidized LDL-regulated microRNAs for evaluating vascular endothelial function: molecular mechanisms and potential biomarker roles in atherosclerosis. *Crit Rev Clin Lab Sci.* 2022;59:40–53.
36. Park YM, Febbraio M, Silverstein RL. CD36 modulates migration of mouse and human macrophages in response to oxidized LDL and may contribute to macrophage trapping in the arterial intima. *J Clin Invest.* 2009;119:136–145.
37. Lankin VZ, Tikhaze AK, Kapel'ko VI, Shepel'kova GS, Shumaev KB, Panasenko OM, Konovalova GG, Belenkov YN. Mechanisms of oxidative modification of low density lipoproteins under conditions of oxidative and carbonyl stress. *Biochemistry.* 2007;72:1081–1090.
38. Chen B, Meng L, Shen T, Gong H, Qi R, Zhao Y, Sun J, Bao L, Zhao G. Thioredoxin attenuates oxidized low-density lipoprotein induced oxidative stress in human umbilical vein endothelial cells by reducing NADPH oxidase activity. *Biochem Biophys Res Commun.* 2017;490:1326–1333.
39. Di L, Maiseyeu A. Low-density lipoprotein nanomedicines: mechanisms of targeting, biology, and theranostic potential. *Drug Deliv.* 2021;28:408–421.
40. León-Pedroza JI, González-Tapia LA, del Olmo-Gil E, Castellanos-Rodríguez D, Escobedo G, González-Chávez A. Low-grade systemic inflammation and the development of metabolic diseases: from the molecular evidence to the clinical practice. *Cir Cir.* 2015;83:543–551.

41. Mei R, Wu M, Ren F. Knockdown of circ_0002194 protects against oxidized low-density lipoprotein-induced cell damage via the regulation of the miR-637/PACS2 axis in human vascular endothelial cells. *Interact Cardiovasc Thorac Surg.* 2022;35:ivac210.
42. Jomova K, Raptova R, Alomar SY, Alwasel SH, Nepovimova E, Kuca K, Valko M. Reactive oxygen species, toxicity, oxidative stress, and antioxidants: chronic diseases and aging. *Arch Toxicol.* 2023;97:2499–2574.
43. Diczfalusy U. Analysis of cholesterol oxidation products in biological samples.*J AOAC Int.* 2004;87:467–473.
44. Federico A, Cardaioli E, Da Pozzo P, Formichi P, Gallus GN, Radi E. Mitochondria, oxidative stress and neurodegeneration. *J Neurol Sci.* 2012;322:2542–2562.
45. Wu J, He Z, Gao X, Wu F, Ding R, Ren Y, Jiang Q, Fan M, Liang C, Wu Z. Oxidized high-density lipoprotein impairs endothelial progenitor cells' function by activation of CD36-MAPK-TSP-1 pathways. *Antioxid Redox Signal.* 2015;22:308–324.

Index

S

T

U

V

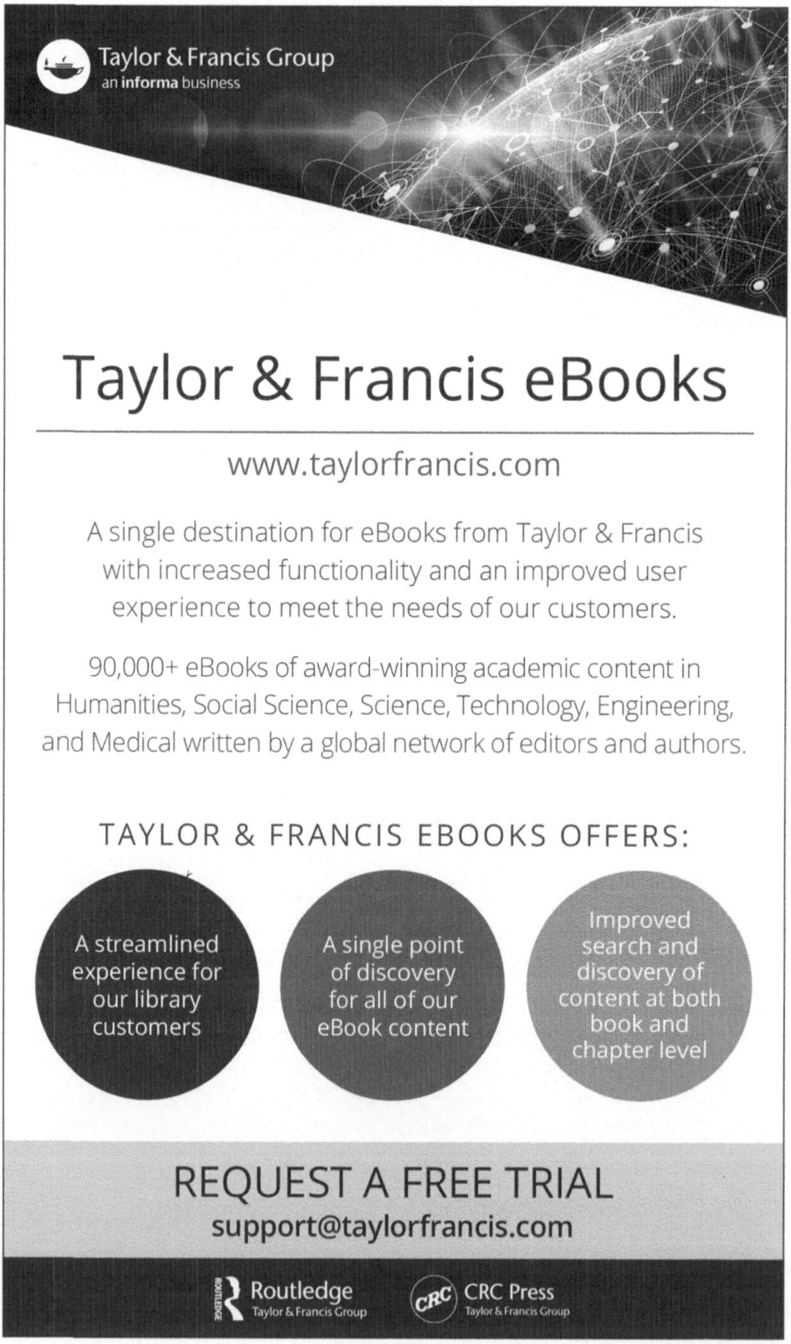
Taylor & Francis Group
an informa business
Taylor & Francis eBooks
www.taylorfrancis.com
A single destination for eBooks from Taylor & Francis with increased functionality and an improved user experience to meet the needs of our customers.
90,000+ eBooks of award-winning academic content in Humanities, Social Science, Science, Technology, Engineering, and Medical written by a global network of editors and authors.
TAYLOR & FRANCIS EBOOKS OFFERS:
A streamlined experience for our library customers
A single point of discovery for all of our eBook content
Improved search and discovery of content at both book and chapter level
REQUEST A FREE TRIAL
support@taylorfrancis.com
Routledge
Taylor & Francis Group
CRC Press
Taylor & Francis Group